WORK THE WORLD WITH JT65 AND JT9

Digital communication
via Amateur Radio!

Steve Ford, WB8IMY

ARRL The national association for AMATEUR RADIO®

Production Staff: **Michelle Bloom, WB1ENT,** Production Supervisor, Layout
Jodi Morin, KA1JPA, Assistant Production Supervisor, Layout
Sue Fagan, KB1OKW, Graphic Design Supervisor, Cover Design
David Pingree, N1NAS, Senior Technical Illustrator

Printed in the USA

ISBN: 978-1-62595-043-7

First Edition

We strive to produce books without errors. Sometimes mistakes do occur, however. When we become aware of problems in our books (other than obvious typographical errors), we post corrections on the ARRL website. If you think you have found an error, please check **www.arrl.org/notes** for corrections. If you don't find a correction there, please let us know by sending e-mail to **pubsfdbk@arrl.org**.

Contents

Foreword

In 1993, Dr Joe Taylor, K1JT, won a Nobel Prize in Physics for his discovery, with Dr Russell Alan Hulse, of the first pulsar in a binary star system. Dr Taylor used the observations of the pulsar to confirm the existence of gravitational radiation in amounts first predicted by Albert Einstein.

You'd think a Nobel Prize, plus several other prestigious awards, would be enough for one lifetime, but Dr Taylor has a passion for Amateur Radio that is every bit as enduring as his love of astrophysics. As a result, his work in weak-signal digital communications has had a substantial impact on the hobby. Dr Taylor's *WSJT* software revolutionized formerly "exotic" activities such as moonbounce and meteor scatter, making them available to virtually all amateurs.

His JT65 and JT9 protocols have done much the same for amateurs who enjoy operating on the HF bands. Thanks to Dr Taylor, hams who are restricted to low profile (or even indoor) antennas and low RF output levels can use JT65 and JT9 to make contacts throughout the world. They can also use JT65 and JT9 to participate in award programs such as ARRL's DX Century Club and Worked All States. In addition, many hams are using JT65 and JT9 to explore the mysteries of propagation, thanks to reverse beacon systems.

If some of what I have described seems like gibberish, don't worry. This book explains it all in clear, concise language. This is your guide to exploring a new world of digital communications. Whether you live in an antenna-restricted environment, or simply want to try something new, the following pages will be your road map to experiencing a fascinating aspect of Amateur Radio.

David Sumner, K1ZZ
Chief Executive Officer
Newington, Connecticut
November 2015

Chapter 1

Getting to Know the "JTs"

The histories of JT65 and JT9 begin not with software, but with a man: Dr Joe Taylor, K1JT. As a Nobel Prize winning scientist who studies pulsars and other distant astronomical objects, Dr Taylor has always had a keen interest in weak signals. When he applied this interest to Amateur Radio in 2001, it sparked a revolution in digital communications.

Dr Joe Taylor, K1JT, speaking at the ARRL Centennial Convention in Hartford, Connecticut in July 2014.

Dr Taylor grappled with a challenge that had dogged wireless communication from the beginning: *free space path loss*. In simple terms, free space path loss is a measure of how much of your transmitted energy is lost between your station and the receiving station. It is proportional to the square of the distance between the transmitter and receiver, and also proportional to the square of the frequency of the radio signal. You don't have to be a mathematician to realize that free space path loss can be enormous.

Throughout the history of Amateur Radio, we've known that we could overcome path loss by either (1) increasing output power, (2) narrowing signal bandwidth (to

PSK31 signals concentrate their energy within a 50-Hz bandwidth.

Since the late 1940s, amateurs have been using the Moon as a giant radio reflector.

concentrate the RF energy), (3) using directional "gain" antennas to focus the transmitted and received signal energy, or a combination of all of these techniques. For example, CW enthusiasts have known for decades that their narrow-bandwidth signals could often be understood when voice communication was all but impossible. When you combine a CW signal with high power levels and gain antennas, it is easy to understand why the mantra "CW gets through when nothing else can" has been true for so many years.

But CW's reign as king of the airwaves ended with the advent of personal computers and Digital Signal Processing, which ushered in the era of modern digital communications. By the end of the 20th century, we saw a proliferation of new digital modes that were often superior to CW when it came to dealing with the impact of free space path loss. For example, a PSK31 signal concentrates most of its transmitted power into a bandwidth of less than 50 Hz, and then uses synchronized changes in the phase of the carrier signal to send information. Other modes use much wider bandwidths, but they work their magic by spreading the information over many separate RF carriers. This approach, known as *redundancy*, allows you to miss some bits here and there while still managing to extract most (or all) of the message. Still other modes use a kind of back-and-forth "handshaking" to make sure each block of data is received error-free before the next block is sent (PACTOR provides a good example).

Dr Taylor's goal was to push the weak-signal communication envelope even further. As a ham, one of his passions involved using the Moon as a giant radio reflector. By bouncing signals off the Moon, it is possible for earthbound stations to communicate on the VHF+ bands over thousands of miles, something that would otherwise be impossible at those frequencies.

Historically, moonbounce could only be achieved with arrays of high-gain antennas like these. The development of JT65 made moonbounce communication feasible for many more amateurs.

This activity, known as *moonbounce* or *Earth-Moon-Earth*, or *EME*, had always been confined to a small circle of hams who could afford 1.5 kW amplifiers for the VHF+ bands, and who had the real estate (and funding) to erect large, high-gain directional antenna arrays. Even with all that power and directivity, being heard via the Moon was a serious challenge. After all, the signal must travel more than a half-*million* miles from the transmitting station to the receiving station. Most moonbouncers back in the day relied on CW; it was the only mode that stood a chance of being discernible after a 500,000 mile trip.

The Birth of JT65

Dr Taylor wondered if it might be possible to create a new mode of digital communication that would be superior to CW in moonbounce applications. Such a mode could provide enormous benefits, making it possible to make moonbounce contacts at much lower RF power levels and with considerably smaller antennas.

After years of research the result was *JT65*, one of the operating modes in Dr Taylor's *WSJT* software suite, which debuted in 2001. With JT65, Dr Taylor pulled out all the digital stops, so to speak. He used redundant multiple-frequency shift keying, data compression and complex *Reed-Solomon* coding. JT65 transmissions are also tightly time synchronized. The transmissions are made at specific intervals and, at the receiving end, the software digs deeply (pardon the oversimplification) to decode the data during those intervals.

JT65 uses one-minute transmit/receive sequences, meaning that you transmit within a one-minute window and then listen during the next minute. Transmission actually begins 1 second after the start of a minute and stops precisely 47.7 seconds later. There is a 1270.5 Hz synchronizing tone and 64 other tones. This combination gives JT65 its unusual musical quality.

Accurate time and frequency synchronization is critical to JT65. Your SSB transceiver needs to reasonably stable, although I've yet to see a modern commercial radio that is too "drifty" for JT65. Drifty computer time is a different matter, however. *Windows* PCs are notorious for loose timekeeping, but there are ways to deal with this, as you'll see.

To communicate information at such weak signal levels, certain sacrifices must be made. One obvious sacrifice is speed. Not only does it take about 5 minutes to complete the average JT65 contact, the highly robust protocol leaves little room for long strings of text in a given transmission. In fact, the longest string of text in JT65 is a mere 13 characters.

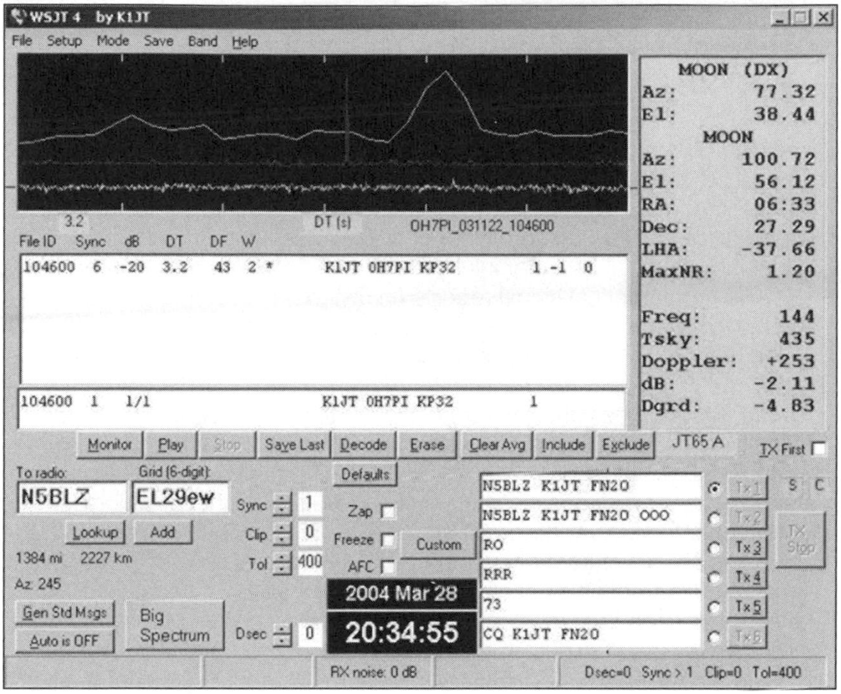

JT65 is one of the modes in K1JT's *WSJT* software suite.

With that in mind, JT65 is not a "conversational" mode like, say, PSK31. Instead, the idea is to exchange just the basic information required for a valid contact: call signs and signal reports.

JT65 worked fabulously well for moonbounce communication. Soon after its debut you heard of hams making EME contacts with as little as 150 W output to single 11-element beam antennas. Moonbounce has never been an activity you'd characterize as "easy," but JT65 suddenly made moonbounce a real possibility for amateurs who never thought they would be capable of such a thing.

Several years passed before some amateurs, including Dr Taylor, began to wonder if JT65 could be a viable technique for earthbound communications, especially on the HF bands where many amateurs were looking for a digital communication mode that could effectively compensate for the absence of high RF power and gain antennas. With antenna restrictions becoming more widespread, hams were eager to find a way to enjoy the hobby on the HF bands without large, highly visible antennas. (Many amateurs are banned from installing outdoor antennas of any kind; their antennas are strictly indoors.)

JT65 and JT9 Protocol Specifics

By Dr Joe Taylor, K1JT

JT65 uses 60 second transmit/receive sequences and carefully structured messages. Standard messages are compressed so that two call signs and a grid locator can be transmitted with just 71 bits. A 72nd bit serves as a flag to indicate that the message consists of arbitrary text (up to 13 characters) instead of call signs and a grid locator. Special formats allow other information such as add-on call sign prefixes (e.g., ZA/K1ABC) or numerical signal reports (in dB) to be substituted for the grid locator. The basic aim is to compress the common messages used for minimally valid QSOs into a minimum fixed number of bits. After compression, a Reed Solomon (63, 12) error-control code converts 72-bit user messages into sequences of 63 six-bit channel symbols.

JT65 requires tight synchronization of time and frequency between transmitting and receiving stations. Each transmission is divided into 126 contiguous time intervals or symbols, each of length 4096/11025 = 0.372 seconds. Within each interval the waveform is a constant-amplitude sinusoid at one of 65 pre-defined frequencies. Frequency steps between intervals are accomplished in a phase-continuous manner. Half of the channel symbols are devoted to a pseudo-random synchronizing vector interleaved with the encoded information symbols. The sync vector allows calibration of time and frequency offsets between transmitter and receiver. A transmission nominally begins at t = 1 s after the start of a UTC minute and finishes at t = 47.8 seconds. The synchronizing tone is at 11025 × 472/4096 = 1270.5 Hz, and is normally sent in each interval having a "1" in the following pseudo-random sequence:

```
100110001111110101000101100100011100111101101111000110101011001
101010100100000011000000011010010110101010011001001000011111111
```

Encoded user information is transmitted during the 63 intervals not used for the sync tone. Each channel symbol generates a tone at frequency 1275.8 + 2.6917 × N × m Hz, where N is the value of the six-bit symbol, $0 \le N \le 63$, and m is 1, 2, or 4 for JT65 submodes A, B, or C. JT65A is the submode always used at HF.

For EME (but, conventionally, not on the HF bands) the signal report OOO is sometimes used instead of numerical signal reports. It is conveyed by reversing sync and data positions in the transmitted sequence. Shorthand messages for RO, RRR, and 73 dispense with the sync vector entirely and use time intervals of 1.486 s (16,384 samples) for pairs of alternating tones. The lower frequency is always 1270.5 Hz, the same as that of the sync tone, and the frequency separation is 26.92 × n × m Hz with n = 2, 3, 4 for the messages RO, RRR, and 73.

JT9 is designed for making minimally valid QSOs at LF, MF, and HF. It uses 72-bit structured messages nearly identical (at the user level) to those in JT65. Error control coding (ECC) uses a strong convolutional code with constraint length K=32, rate r=1/2, and a zero tail, leading to an encoded message length of (72+31) × 2 = 206 information-carrying bits. Modulation is nine-tone frequency-shift keying, 9-FSK. Eight tones are used for data, one for synchronization. Eight data tones means that three data bits are conveyed by each transmitted information symbol. Sixteen symbol intervals are devoted to synchronization, so a transmission requires a total of 206 / 3 + 16 = 85 (rounded up) channel symbols. The sync symbols are those numbered 1, 2, 5, 10, 16, 23, 33, 35, 51, 52, 55, 60, 66, 73, 83, and 85 in the transmitted sequence.

Each symbol lasts for 6912 sample intervals at 12000 samples per second, or about 0.576 seconds. Tone spacing of the 9-FSK modulation is 12000/6912 = 1.736 Hz, the inverse of the symbol duration. The total occupied bandwidth is 9 × 1.736 = 15.6 Hz.

The most striking difference between JT65 and JT9 is the much smaller occupied bandwidth of JT9: 15.6 Hz, compared with 177.6 Hz for JT65A. Transmissions in the two modes are essentially the same length, and both modes use exactly 72 bits to carry message information. At the user level the two modes support nearly identical message structures.

JT65 signal reports are constrained to the range –1 to –30 dB. This range is more than adequate for EME purposes, but not really enough for optimum use at HF and below. S/N values displayed by the JT65 decoder are clamped at an upper limit –1 dB. Moreover, the S/N scale in present JT65 decoders is nonlinear above –10 dB.

By comparison, JT9 allows for signal reports in the range –50 to +49 dB. It manages this by taking over a small portion of "message space" that would otherwise be used for grid locators within 1 degree of the South Pole. The S/N scale of the present JT9 decoder is reasonably linear (although it's not intended to be a precision measurement tool).

With clean signals and a clean nose background, JT65 achieves nearly 100% decoding down to S/N = –22 dB and about 50% at –24 dB. JT9 is about 2 dB better, achieving 50% decoding at about –26 dB. Both modes produce extremely low false-decode rates.

Early experience suggests that under most HF propagation conditions the two modes have comparable reliability. The tone spacing of JT9 is about two-thirds that of JT65, so in some disturbed ionospheric conditions in the higher portion of the HF spectrum, JT65 may perform better.

JT9 is an order of magnitude better in spectral efficiency. On a busy HF band, the conventional 2-kHz-wide JT65 sub-band is often filled with overlapping signals. Ten times as many JT9 signals can fit into the same frequency range, without collisions.

JT65 signals often decode correctly even when they overlap. Such behavior is much less likely with JT9 signals, which fill their occupied bandwidth more densely. JT65 may also be more forgiving of small frequency drifts.

JT65 provided the answer – specifically with a variant of the protocol known as *JT65A,* which is part of Dr Taylor's *WSJT-X* software suite. (Throughout this book I will refer to JT65A simply as "JT65.")

With JT65 amateurs suddenly discovered that they could work the world on the HF bands with just a couple of watts and indoor antennas. They found themselves sending and receiving JT65 communications under conditions that would present severe challenges to modes such as CW and PSK31 – even with outdoor antennas and higher power levels.

In addition to Dr Taylor's work, other software developers created dedicated JT65 applications (mostly for Microsoft *Windows* or *Linux* operating systems) that greatly streamlined the process of making successful JT65 contacts. As a result, JT65 activity blossomed on the HF bands. In the year this book was published, JT65 had become one of the most popular HF digital modes, second only to PSK31.

JT9

As any dedicated scientist will tell you, the quest for discovery never ceases. Dr Taylor wasn't content to simply sit back and tweak his existing programs. Over time he added new modes to *WSJT-X*, including another that quickly captured the attention of the HF digital community: *JT9*.

Dr Taylor developed JT9 specifically for use on the LF, MF and low HF bands. Its coding is similar to JT65 (although JT9 employs a 9-FSK modulation technique) and it also uses 1-minute transmit/receive intervals. JT9 is even more sensitive than JT65, and there are also versions that use transmission intervals of 2, 5, 10 and even 30 minutes. The longer transmission intervals make it possible to communicate at extremely weak signal levels, far weaker than would ever be possible with JT65.

JT9 is not as widely used as JT65, but its popularity is increasing. When this book was written, JT9 was only available in Dr Taylor's *WSJT-X* software, although I suspect it will appear in other software applications in due time. Also, in 2015 the FCC announced that it would soon grant amateurs access to new bands at 630 and 2200 meters. Preliminary tests have shown great promise for the use of JT9 on these bands.

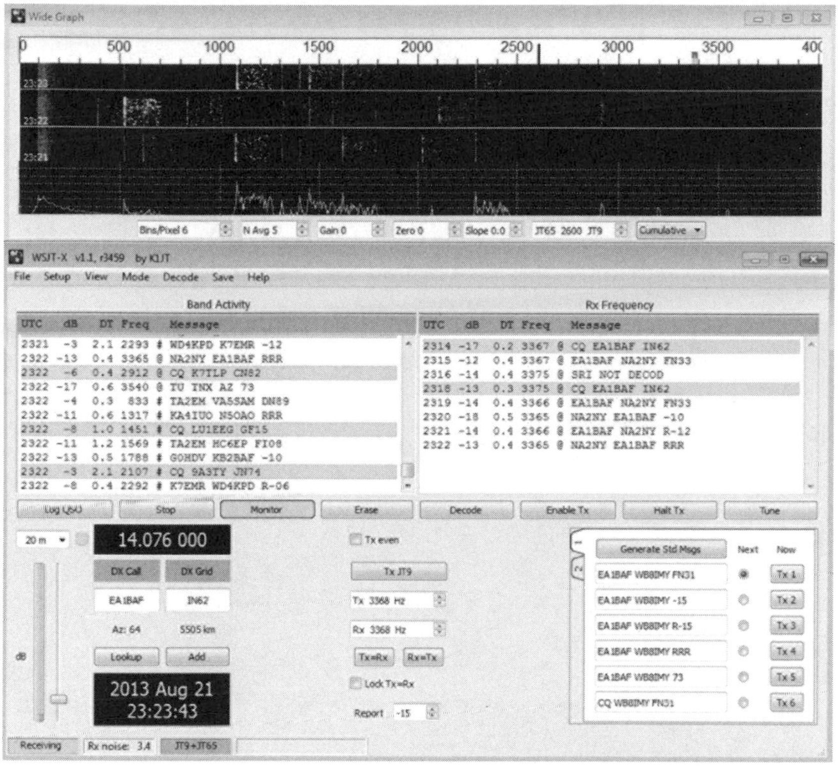

Thanks to the development of *WSJT-X*, amateurs can use JT65 and JT9 on the HF bands.

What is the Attraction?

It is fair to ask why so many amateurs would be attracted to communication modes that don't allow true conversations. As mentioned previously, JT65 and JT9 are designed to exchange just the basic facts – the minimum information necessary to constitute a valid contact for the purposes of various awards such as the ARRL DX Century Club (DXCC).

A large part of the answer may be cultural.

Once upon a time, enjoying a wireless conversation was not a common occurrence, especially when distances between participants were measured in thousands of miles. Back then, an Amateur Radio operator was seen as a unique individual with the knowledge and equipment necessary to do what only commercial telephone networks could achieve.

The communication that resulted was equally unique. Hams on other continents found themselves using radio technology to enjoy international

JT65 and JT9 make it possible for amateurs using indoor antennas and low power levels to earn prestigious awards such as the ARRL DX Century Club.

conversations at a time when such exchanges were often limited to governments or wealthy individuals. In those days, Amateur Radio was the outlier, the odd exception to the rule. To many, hams were wizards who worked with strange and potentially dangerous tools. The fact that these amateurs cast their homemade radio signals around the world with impunity was almost unbelievable.

We live in a very different world today. Global communication is inexpensive and common, and does not require specialized knowledge or skill. With worldwide wireless being an accepted fact of life, the allure of Amateur Radio as a means of enjoying long-distance chats has faded considerably. As a result, we've witnessed declining interest in casual conversation among amateurs. Instead, modern hams often see radio more as a means to an end: earning an award, studying propagation, learning about technology, competing in on-air contests, testing software, serving the community, or enjoying the satisfaction of building or restoring a piece of equipment.

JT65 and JT9 are ideal for this new reality. Not only do they allow you to make contacts with minimal antennas and low RF output power, each completed contact counts for DXCC and many other awards. As you'll learn later in this book, JT65 and JT9 are also excellent tools for studying the effects of propagation, or testing new radios and antennas.

Because of the lengthy transmission and reception times, JT65 and JT9 are also rather relaxing. If you're careful, it is possible to hold a conversation with a friend or family member at your station while still completing a contact. This isn't to say that these modes are not without their challenges, though. Being "careful" means that you'll still need to pay close attention to your computer monitor and clock, as you will soon discover. JT65 and JT9 contacts can also occasionally become what I would call "non-routine," and that tends to keep you on your toes as well.

JT65 and JT9 contacts are not mindless exchanges of data. The software is only a tool to encode and decode information. *You* are still very much in control. Abbreviated as they may be, JT65 and JT9 contacts are still human-to-human. The text displayed on your screen is sent at the command of another individual, perhaps on the opposite side of the world. And just like you, he or she is staring eagerly at their monitor, awaiting your response!

Chapter 2

Let's Build a JT-Capable Station

If you are already active with HF digital with modes such as PSK31, chances are you can skip much of this chapter. The hardware you've already installed will work perfectly well for JT65 and JT9.

However, if you must skip ahead, be sure to read the **Software** discussion in this chapter.

The HF Transceiver

The radio requirements for JT65 and JT9 are surprisingly straightforward. You don't need to purchase a special radio with sophisticated features. All you need is an SSB voice transceiver.

Even an old SSB transceiver, such as this Drake TR-4, can be used with JT65 or JT9 if you allow time for the radio to completely warm up and stabilize.

Any transceiver made within the last 20 years should work well. Even very old SSB rigs can be pressed into service. When considering an older radio, however, keep frequency stability in mind. Older transceivers, especially radios that use vacuum tubes, may tend to drift in frequency. Drift is deadly to digital communications. If you must use an old rig as your HF digital transceiver, you may need to allow it to warm up for as long as 30 minutes prior to operating.

If your radio dates from about 1990 to the present day, you'll likely be in fine shape. These rigs are stable and most include the features you'll need for HF digital. The singular exception involves some of the inexpensive SSB transceivers designed for low-power (QRP) operating. These rigs aren't usually intended for digital applications, so their stability may be questionable. They may also not offer connections to external devices through jacks known as *ports*.

Radio Ports

Modern radios get along surprisingly well with computers and other external devices. In fact, most modern HF transceivers are designed with external devices in mind. They offer a variety of ports to communicate with the outside world.

Nearly every HF rig manufactured within the past decade includes an "accessory" port of some kind. Typically these are multipin jacks (as many as 13 pins) that provide connections for audio into and out of the radio, as well as a pin that causes the radio to switch from receive to transmit whenever the pin is grounded. This is often called the *PTT* or *Push to Talk* line. Some manufacturers also call it the "Send" line. Take a look at the typical accessory port shown in **Figure 2.1**.

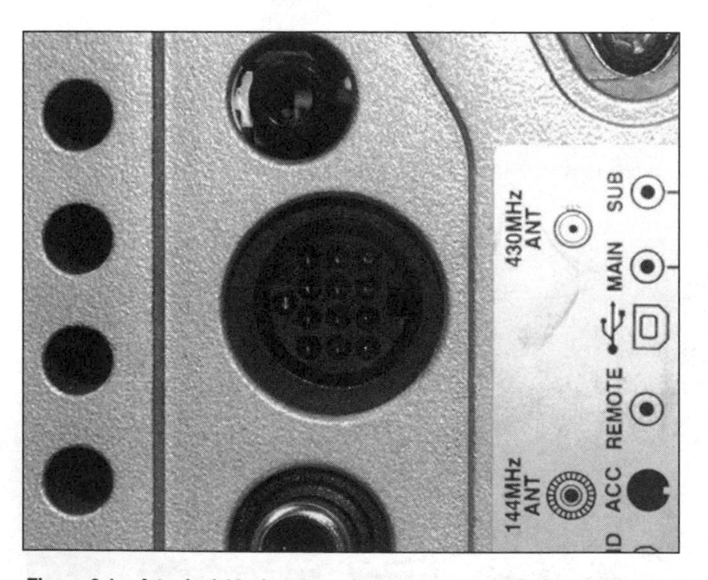

Figure 2.1 – A typical 13 pin transceiver accessory jack. Among the pins of most interest to HF digital operators are those that carry receive audio, transmit audio and the PTT (Push To Talk) line.

These accessory ports are ideal for connecting the kinds of interface devices we use to operate HF digital. In addition to the PTT function, accessory ports provide receive audio output at *fixed* levels that never change no matter where the VOLUME knob is set. This is a highly convenient feature that you'll appreciate when operating late at night after everyone has gone to bed. You can turn the VOLUME knob to zero and still have all the receive audio you need.

Be aware that some radio manufacturers label accessory ports as "data" or "digital" ports. This causes no end of confusion because modern rigs often offer two types of connections: a true accessory port with audio and transmit/receive keying lines, and another port that allows a computer to actually control the radio. The confusion occurs when hams attempt to figure out which ports they need to use.

For the purpose of getting on the air with JT65 or JT9, the type of radio "control" we care most about is the ability to switch from receive to transmit and back again. That connection is available at the *accessory port*, even though the port may go by a different name.

On the other hand, the kind of control the transceiver manufacturers often have in mind goes way beyond the act of simply switching between transmit and receive. They are talking about computer control of almost every function of the radio; that's a different animal entirely. Full computer control usually involves software that does many things such as displaying and changing the transceiver's frequency, raising and lowering the audio level, scanning memory frequencies and a great deal more. Some computer control software is so elaborate that the radio itself can be placed out of sight and all control conducted at the keyboard and monitor screen. Many amateurs use this capability to control their rigs remotely over the Internet at great distances.

Transceivers have separate ports for this type of computer interfacing and these are most definitely are *not* accessory ports. On the contrary, they are ports strictly designed to swap data with external computers. They come in several varieties …

TTL: Transistor-Transistor Logic. These ports require a special interface to translate the serial communication from your computer to TTL pulses that your radio can comprehend.

USB: Universal Serial Bus. Although the consumer electronics world adopted USB years ago, transceiver manufacturers have been somewhat slower to catch on.

RS232: This is a serial port that can be connected directly to your computer if your computer has a serial (COM) port. Most modern computers have done away with serial ports, but you can use a USB-to-serial converter to bridge the gap.

Ethernet: This port allows the transceiver to become a "network device," just like your wireless router, printer, etc. Only a handful of transceivers offer Ethernet ports at this time.

Most JT65 and JT9 software packages include the ability to directly read the frequency of your radio through these ports. This is handy if your software performs logging functions, or if it sends activity reports to the Internet (more about this in the next chapter and beyond). Neither function is essential to your enjoyment of JT65 or JT9, however.

A 9-pin male RS232 serial port.

No Radio Ports?

What if your SSB transceiver doesn't have an accessory port? No problem.

You can use the microphone jack as your connection for transmit/receive switching as well as the audio input. For the audio output, you can use the external speaker or headphone jack. This isn't an ideal situation, but it works. In fact, many HF digital operators take this approach.

Duty Cycle

When talking about your transceiver, we need to spend a little time discussing the concept of *duty cycle*. A somewhat crude definition of duty cycle is the percentage of the total transmitting time during which the radio is producing RF output. In HF digital terms, think of duty cycle as the amount of time your radio is generating RF during any given transmission compared to the amount of the time during the same transmission when RF output falls to zero. A 100% duty cycle would mean that your radio is cranking out RF continuously throughout the entire transmission; the RF output level never falls to zero.

Now you may say, "I guess I'm always operating at a 100% duty cycle. After all, my radio is always generating RF whenever I'm transmitting."

Not necessarily.

When you are transmitting digital, CW or even SSB voice, your rig may not be operating at a 100% duty cycle. Consider SSB voice as an example. Whenever you speak into the microphone, the RF output level changes dramatically as your voice changes. It can go from 100% output to zero in a fraction of a second. The same is true for CW. Whenever your CW key is open between the dots and dashes, your transceiver output is zero.

Measured over a period of time (your transmitting time), the duty cycle of SSB voice is actually about 40%; CW is often as low as 30% or even less if you are a particularly slow sender. When it comes to JT65 and JT9, they will push your radio to a duty cycle of nearly 100%.

But why should you care?

The answer is that your radio may not be designed for the type of punishment a high duty cycle transmission can inflict. When you operate your radio at a 100% duty cycle, you are demanding that its final amplifier circuits produce the full measure of output – whatever you've set that output level to be – for the entire time you are transmitting. The result is heat, and potentially a lot of it. Apply enough heat to a circuit for a sufficient length of time and you'll see components begin to fail, sometimes in spectacular fashion.

If you have a 100 W transceiver, the good news is that you'll almost never need to operate it at full output for JT65 and JT9. In fact, 100 W is an obscene amount of RF power for these modes. Fifty watts of RF is considered "high power" in the JT65/JT9 world; most operators use much less. On that basis, your 100 W transceiver should be able to handle JT65 or JT9 at 50 W or less without breaking a sweat.

On the other hand, if you are operating with a radio designed for only 5 or 10 W output and you are running it at full throttle, take care. If you notice that your radio is becoming particularly hot, reduce the RF output. You may be pushing the rig to the edge of its design limits.

Computers

From the early 1980s through about 2005, the desktop computer was king among ordinary consumers and Amateur Radio operators. This is a computer in a separate, stand-alone case connected to a monitor screen, keyboard and mouse. Inside the computer case there is a sound device of some sort, either a dedicated *sound card* or a set of sound-processing chips on the motherboard. The computer connects to peripheral devices through the use of serial (COM) ports.

From 2005 onward we saw two important changes. One was the fact that laptop computers became more powerful and affordable. The other

At the time this book was written, desktop PCs were fading in the consumer world. Among hams, however, they were still widely used.

was that serial ports disappeared in favor of USB ports in both laptops *and* desktop machines.

By 2010 laptops began to dominate Amateur Radio stations. As hams upgraded their computers, they no longer saw the need for bulky desktop systems when sleek laptops would do quite nicely.

As this book went to press, even laptops were facing stiff competition from *tablet* computers such as the Apple iPad. In Amateur Radio stations, laptops and desktops are still the most popular computers, but this is likely to change as more ham applications are developed for tablets.

For now, however, our focus will remain primarily on laptops and desktops. With that in mind, what kind of laptop or desktop do you need for JT65/JT9 operating?

The good news is that an ordinary off-the-shelf consumer-grade computer will likely be sufficient. If you are buying new, don't overspend for a powerful computer you won't need. Frankly, a decent used computer that's only a couple of years old will be adequate.

In terms of operating systems, the vast majority of JT65/JT9 software is written for Microsoft *Windows*. However, there are versions of Dr Taylor's *WSJT-X* software available for *Linux* and Mac computers.

If you are considering a new or used computer (desktop or laptop), here are a few rule-of-thumb shopping specifications:

• The processor clock frequency ("speed") should be 1.5 GHz or better.

• The computer should have as much memory as possible, not just to run the ham applications smoothly, but the operating system as well. For *Windows 7, 8, or 10*, I'd recommend at least 8 GB of RAM; more is always better.

• Either a built-in wireless (Wi-Fi) modem or an Ethernet port. Although it isn't necessary for ham work, chances are you'll want to connect your station to the Internet from time to time. If so, you'll need a wireless modem or Ethernet port to do so.

• A CD-ROM drive for loading new software that is only available on CD (a less common need today, but don't sell yourself short).

The Importance of Sound

With few exceptions, every computer you are likely to purchase today will include a either a dedicated sound card or a sound chipset. This feature is absolutely critical for HF digital operation because most of the modes you'll enjoy – including JT65 and JT9 – depend on sound devices to act as radio *modems* – *mod*ulators/*dem*odulators.

The audio from your radio enters your computer via the sound device where it is converted (demodulated) to digital data for processing by your software. The results are words or images on your computer monitor. When you want to transmit, this same sound device takes the data from your software, such as the words you are typing, and converts it to shifting audio tones according to whatever mode you are using. This conversion is a form of modulation. The tones are then applied to your radio for transmission.

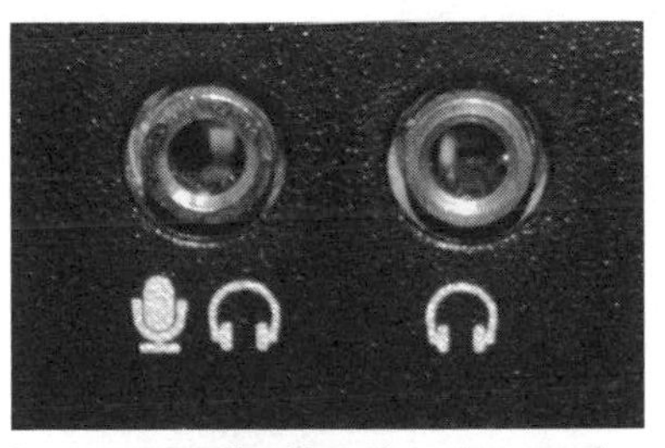

Laptops usually offer microphone and headphone audio jacks. These are fine for HF digital use.

The simplest built-in sound devices are those found in laptops and tablets. They provide two ports to the outside world: microphone (audio input) and headphone (audio output). These are perfectly adequate for JT65/JT9 work. Desktop computers often have a similar arrangement, although the output port is usually labeled "speaker." A "line" input may also be included for stronger audio signals. In all cases these ports come in the form of 1/8-inch stereo jacks.

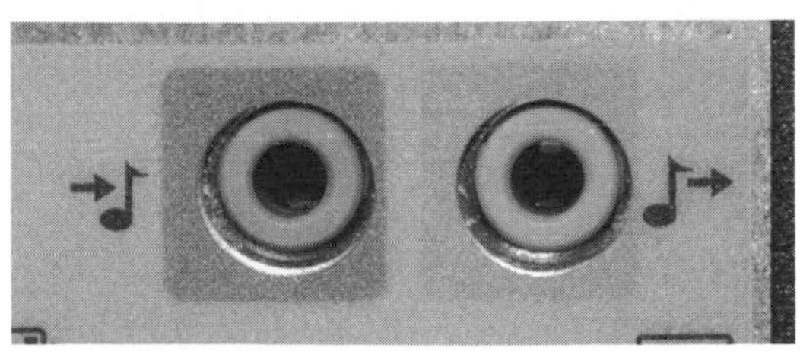

Rear panel motherboard audio ports in a typical desktop computer.

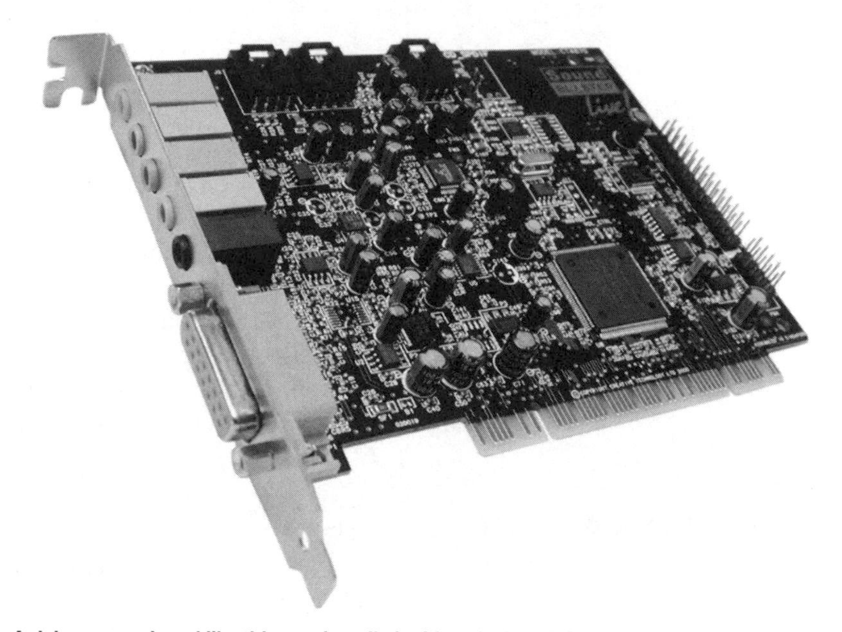

A deluxe sound card like this one installs inside a desktop PC. Note all the input and output ports.

Some desktops computers offer sound cards that plug into the motherboard. These devices are more elaborate. Some sound cards can offer as many as 12 external connections. At the rear of your computer you may find LINE IN, MIC IN, LINE OUT, SPEAKER OUT, PCM OUT, PCM IN, JOYSTICK, FIREWIRE, S/PDIF, REAR CHANNELS or SURROUND jacks, just to name a few. For JT65/JT9 use, the important jacks are MIC or LINE IN and SPEAKER OUT.

Later in this chapter we'll discuss how to connect these sound devices to your radio, but one item needs to be briefly mentioned now: the *interface*. As you'll see later, the interface is yet another critical component because it is the link between the computer and the radio. The reason to mention it now is because the trend in interface technology has been to incorporate the sound device into the interface itself. At the time of this writing there are several interfaces by companies such as microHAM, Timewave, West Mountain Radio and TigerTronics that feature their own built-in sound devices. These interfaces are extremely convenient because they work independently of whatever sound device you have in your computer. Just plug in their USB cables and you're good to go. It doesn't matter what kind of computer you are using; the interface will work with it. You also avoid a rat's nest of wiring between the computer and the radio.

Software

When we're talking about putting together a JT65/JT9 station, a brief word about software is in order.

As of 2015, the vast majority of amateurs were running *Windows 7* or *8* operating systems on their station computers. However, Microsoft introduced *Windows 10* this year and it is likely that by 2017 most *Windows* users will be running *Windows 10*. Fortunately, all of the popular JT65/JT9 software that runs under *Windows 7* and *8* also runs quite well under *Windows 10*.

Apple Macintosh owners are obviously quite fond of the various *MacOS* incarnations, which have proven themselves to be efficient and reliable operating systems. The universe of Mac users is growing, but Amateur Radio software titles for the Mac are still few in number. At the time this book was written, the only Mac-compatible software for JT65/JT9 was *WSJT-X*.

The *Linux* operating system in its various distributions enjoys a loyal following among amateurs who like writing their own software and tinkering at the operating system level. As with the Mac, *WSJT-X* is presently the only JT65/JT9-compatible software available.

The three operating systems have their vocal advocates and I'd be a fool to champion one over another. All have their advantages and disadvantages to consider from an Amateur Radio point of view.

Windows

Pro: Sheer variety. Most Amateur Radio software is written for *Windows* so in addition to your JT65/JT9 applications, you have a rich selection of other compatible software to choose from.

Con: Because of the widespread use of *Windows*, it is a favorite target for hackers. Anti-virus software is a must and this can significantly hamper the efficiency of *Windows* and cause other annoying issues. Also, *Windows* can be expensive if purchased and installed separately.

MacOS

Pro: Stability and performance. Highly intuitive and easy to use. Also, hackers have only rarely targeted *MacOS*.

Con: *MacOS* runs only on Apple Macintosh computers. It can be made to run on PCs, but it isn't an exercise for the fainthearted. Amateur Radio software for *MacOS* is limited.

Linux

Pro: Open source and free of charge. Depending on the version, it can be quite efficient and powerful. Hackers generally don't bother designing viruses for *Linux*. They prefer easier prey.

Con: The directory structure and commands may be very different

At the time this book was written, only Dr Taylor's *WSJT-X* software offered JT9.

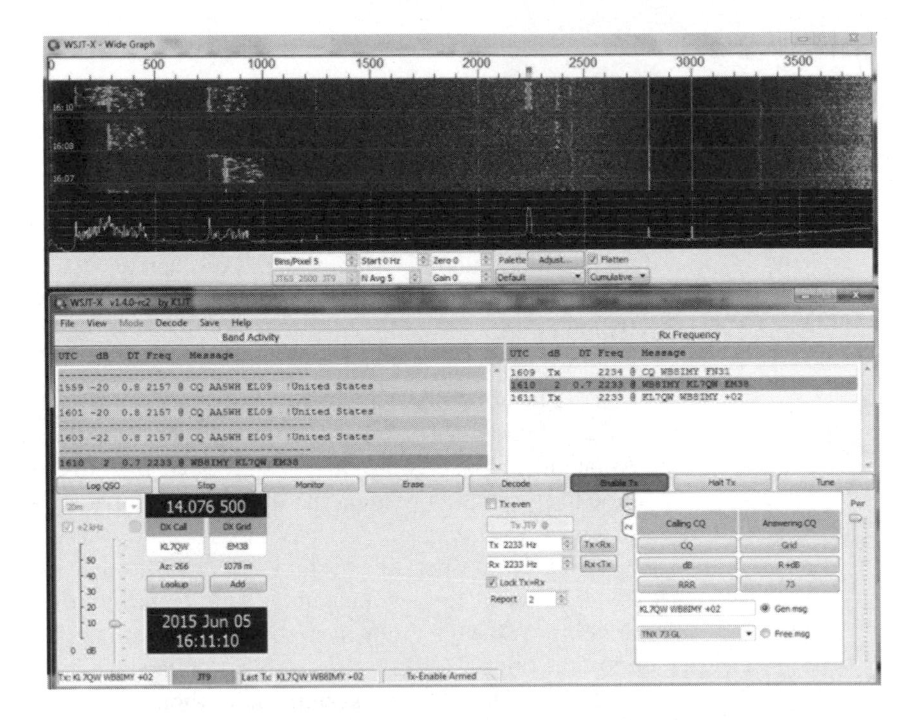

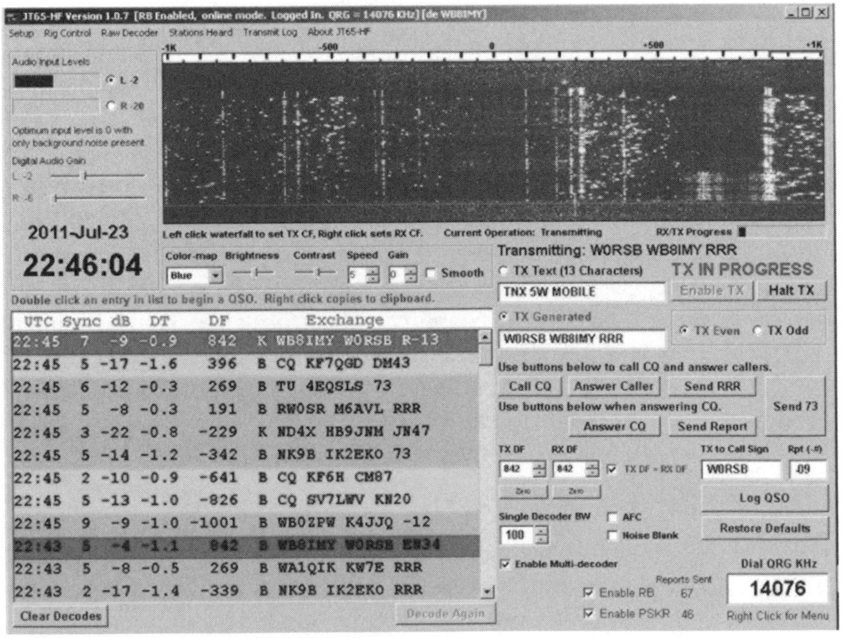

JT65-HF by Joe Large, W6CQZ. Even though this program is no longer supported, it remains very popular among JT65 operators.

JT65-HF Version 1.0.7 [RB Enabled, online mode. Logged In. QRG = 14076 KHz] [de WB8IMY]

Setup Rig Control Raw Decoder Stations Heard Transmit Log About JT65-HF

-1K -500 0 +500 +1K

Audio Input Levels
L -2
R -20
Optimum input level is 0 with only background noise present.
Digital Audio Gain
L -2
R -6

2011-Jul-23
22:46:04

Left click waterfall to set TX CF, Right click sets RX CF. Current Operation: Transmitting RX/TX Progress

Color-map Brightness Contrast Speed Gain
Blue 5 0 Smooth

Double click an entry in list to begin a QSO. Right click copies to clipboard.

UTC	Sync	dB	DT	DF		Exchange
22:45	7	-9	-0.9	842	K	WB8IMY W0RSB R-13
22:45	5	-17	-1.6	396	B	CQ KF7QGD DM43
22:45	6	-12	-0.3	269	B	TU 4EQSLS 73
22:45	5	-8	-0.3	191	B	RW0SR M6AVL RRR
22:45	3	-22	-0.8	-229	K	ND4X HB9JNM JN47
22:45	5	-14	-1.2	-342	B	NK9B IK2EKO 73
22:45	2	-10	-0.9	-641	B	CQ KF6H CM87
22:45	5	-13	-1.0	-826	B	CQ SV7LWV KN20
22:45	9	-9	-1.0	-1001	B	WB0ZPW K4JJQ -12
22:43	5	-4	-1.1	842	B	WB8IMY W0RSB EN34
22:43	5	-8	-0.5	269	B	WA1QIK KW7E RRR
22:43	2	-17	-1.4	-339	B	NK9B IK2EKO RRR

Clear Decodes Decode Again

Transmitting: W0RSB WB8IMY RRR
TX Text (13 Characters) TX IN PROGRESS
TNX 5W MOBILE Enable TX Halt TX
TX Generated
W0RSB WB8IMY RRR TX Even TX Odd
Use buttons below to call CQ and answer callers.
Call CQ Answer Caller Send RRR Send 73
Use buttons below when answering CQ.
Answer CQ Send Report
TX DF RX DF TX to Call Sign Rpt (-#)
842 842 TX DF = RX DF W0RSB 09
Zero Zero Log QSO
Single Decoder BW AFC
100 Noise Blank Restore Defaults
Enable Multi-decoder Dial QRG KHz
Reports Sent
Enable RB 67 14076
Enable PSKR 46 Right Click for Menu

compared to *Windows*. Amateur Radio software selection is very limited.

Now that we have covered the computer operating systems, here are your JT65/JT9 software choices (at least as of press time):

• *WSJT-X* by Dr Joe Taylor, K1JT: Available free for *Windows*, *Linux* and Mac computers at **http://physics.princeton.edu/pulsar/K1JT/wsjtx.html**.

• *JT65-HF* by Joe Large, W6CQZ: Available free for *Windows* only and for JT65 only at **http://jt65-hf.com/downloads/**. It is also available at **www.arrl.org/hf-digital** under the "Resources" tab. It is important to mention that W6CQZ has ceased supporting *JT65-HF*. However, it remains one of the most popular programs for JT65 operating. Other amateurs have created variations of *JT65-HF* that you may want to investigate. *JT65-HF-Comfort* is available at **www.funkamateure-dresden-ov-s06.de/index.php?article_id=315**. The *HB9HQX edition* of JT65-HF can be downloaded at **http://sourceforge.net/projects/jt65hfhb9hqxedi/**.

• *Ham Radio Deluxe*: This popular multimode software added JT65 in 2015. You'll find it at **www.ham-radio-deluxe.com**.

The Interface – the Full Story

If you're like most amateurs, you already own most of the components we've discussed so far. You probably have an HF transceiver and, like 90% of most US households, you have a computer in residence. The one piece of hardware that you may not own is the one item that brings these components together to create a JT65/JT9 station: the *interface*.

At rock bottom an interface has only one job to do: to allow the computer to toggle the radio between transmit and receive. It achieves this by using a signal from the computer to switch on a transistor (see **Figure 2.2**). This transistor "conducts" and effectively brings the transceiver's *PTT* (Push to Talk) line to ground potential or very close to it. When the

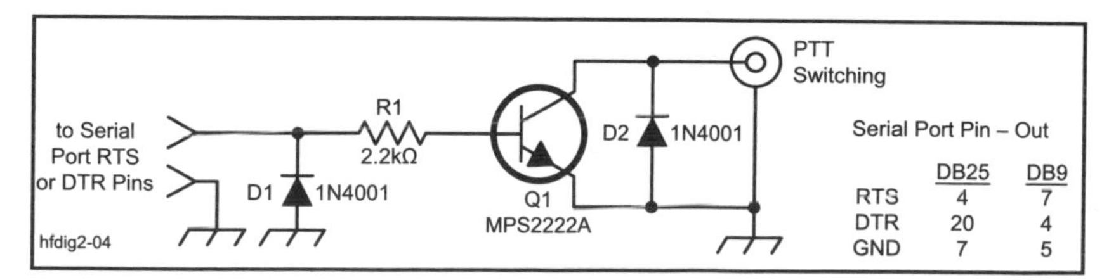

Figure 2.2 – The simplest way to key a radio using a computer is through a single-transistor circuit like this. The input connects to the serial cable coming from the computer (or from the USB/serial adapter). Either the RTS or DTR pins can be used, depending on what your software requires. The pin numbers for 25- and 9-pin plugs are shown.

PTT line is grounded, the transceiver switches to transmit. When the signal from the computer disappears, the transistor no longer conducts and the PTT line is electrically elevated above ground. The result is that the transceiver returns to the receive mode.

The signal from the computer appears at a specific pin on a serial (COM) or USB port. Your JT65/JT9 software generates the signal when it is time to transmit.

If an interface can be so straightforward, couldn't you just build your own? Yes, you could. Many amateurs enjoy JT65/JT9 with simple interfaces like the one shown here. In addition to the switching circuit in Figure 2.2, they connect shielded audio cables between the computer and the radio to carry the transmit and receive audio signals. See **Figure 2.3**.

In Figure 2.3 you'll note that the audio lines include 1:1 isolation transformers. The reason for this is to avoid the dreaded *ground loop*. A ground loop results when current flows in conductors connecting two devices at different electrical potentials. In your HF station the conductors in question are usually the audio cables running between the radio and the computer.

A ground loop typically manifests itself as a hum that you'll hear in your receive audio, or that other stations will hear in your transmit audio.

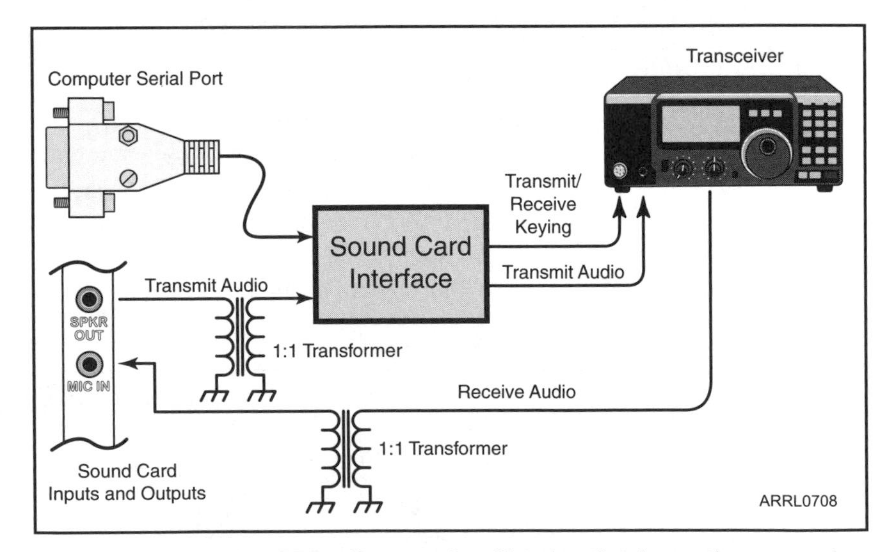

Figure 2.3 – A connection diagram for a sound card interface that does *not* have a sound device built in. If your computer lacks a serial port, you will need to purchase a USB/serial adapter cable. If your chosen interface *does* have a sound device built in, the only connection to your computer will be the through the interface's USB cable; the computer sound card is not used at all and receive audio goes directly to the interface.

The hum can be so loud it will distort the received or transmitting signals, making digital communication impossible. The isolation transformers effectively break the ground loop path while still allowing the audio signals to reach their destinations.

To Roll or Not to Roll

There are good reasons to roll your own interface, the cost savings being chief among them. On the other hand, if you purchase an interface off the shelf you'll be able to benefit from enhanced design features, depending on how much you want to pay. The short list of useful features includes …

• **Independent Audio Level Controls.** These are knobs on the front panel of the interface that allow you to quickly raise or lower the transmit or receive audio levels. Many amateurs prefer to manage the audio levels in this fashion compared to doing it in software.

• **CW keying.** Full-featured interfaces handle more than just JT65/JT9. They can also use keying signals from the computer to send Morse code with a separate connection to the transceiver's CW key jack. This allows you to send CW from your keyboard rather than with a hand key, a useful feature for higher speed CW exchanges during contests.

• **Microphone input.** If you are making JT65/JT9 connections to your radio through the microphone jack rather than the rear panel accessory port, you'll need to unplug the interface cable whenever you want to use your microphone for a voice conversation. To make operating more convenient, some interfaces allow you to keep your microphone plugged into the interface at all times, switching between your microphone or computer as necessary.

• **Transceiver control.** Remember that a basic interface does not allow your computer to truly control your radio, except in the sense that it can switch your radio between transmit and receive. Deluxe interfaces include the extra circuitry needed to allow full computer control of your transceiver. Sometimes referred to as *CAT* (Computer Aided Transceiver), this is a separate function that passes all the available controls from your radio to your computer. Depending on the type of transceiver you own and the software you are using, CAT allows you to change frequency, raise and lower power levels and much more. If your transceiver has the ability to connect directly to your computer through an RS-232 serial connection, USB cable or Ethernet port, you don't need the CAT feature. The CAT function is primarily intended for radios that use transistor-transistor (TTL) signals for control. Manufacturers sell their own CAT interfaces,

The microHAM USB III interface with a built-in sound device.

The MFJ-1275 interface.

The TigerTronics SignaLink USB interface also includes a built-in sound device.

The Timewave Navigator interface with audio level controls for its internal sound device.

The RigBlaster Advantage interface from West Mountain Radio has a built-in sound device.

but they tend to be expensive. An interface with CAT functionality brings everything together in one affordable box.

• **Built-in Sound Device.** As we discussed earlier, several interface designs include a built-in sound device. This is particularly handy in that it liberates the sound device in your computer for other functions. You can enjoy music on your computer, for example, without having to worry that you are meddling with the sound levels you've set up for JT65/JT9 operating. In addition, an interface with a built-in sound device greatly reduces the number of cables connecting the computer and radio. The audio signals, as well as transmit/receive keying functions, are all carried over a single USB cable; there are no connections to your computer's sound ports.

• **Pre-Made Cables.** Most commercial interface manufacturers either include cables specifically wired for your radio free of charge, or offer them at an additional cost. This significantly reduces the hassle of wiring your JT65/JT9 station.

At the time of this writing, a basic off-the-shelf interface costs about $50; a multi-featured deluxe interface runs as high as $400. You'll need to shop among the manufacturers to find an interface that has the features you desire at a cost you are willing to pay. The most popular interface manufacturers include . . .

microHAM: **www.microham-usa.com**
MFJ: **www.mfjenterprises.com**
TigerTronics: **www.tigertronics.com**
West Mountain Radio: **www.westmountainradio.com**
Timewave: **timewave.com/**

Putting it all Together

If you've chosen an interface with a built-in sound device, assembling your station is relatively easy. You'll need a set of audio and PTT cables to connect the interface to your transceiver, either at the microphone and headphone jacks, or at the accessory jack. As I've already mentioned, you can purchase this cable from the interface manufacturer or make your own. The USB cable from the interface simply plugs into your computer.

The USB connection to your computer can be a little tricky in one respect, though. When you plug the USB cable into the computer for the first time, the computer may attempt to load and run a *driver* application so it can "talk" to your interface. This driver may already exist on your computer, or you may need to load it from a CD supplied by the manufacturer. Once the driver is loaded, the computer will recognize the interface every time you plug it in thereafter.

Even though the interface is connecting to your computer through a USB cable, it is depending on good old fashioned serial communication just as though the connection had been made through a COM port. The interface accomplishes this by creating a *virtual COM port* in your computer. In other words, it uses software to emulate the function of a COM port.

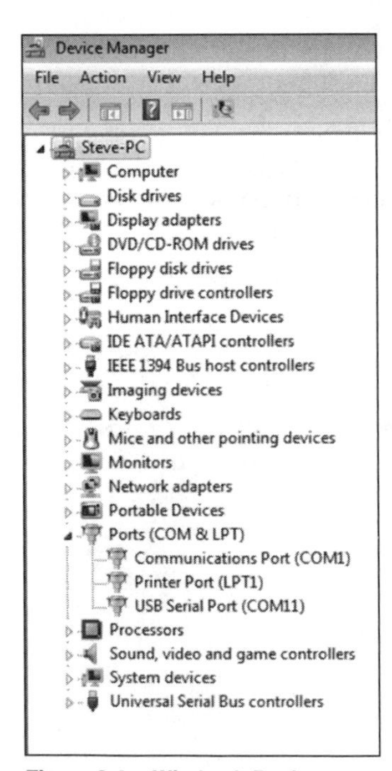

Figure 2.4 – *Window's* Device Manager showing the available ports. Note the "USB Serial Port" is labeled "COM 11."

Why is it important for you to know this? The answer is that your digital software will need to be configured so that it "knows" which COM port to use for PTT keying, CAT functions, etc. That means you'll need to know this as well!

In *Windows* it is a matter of going to the **Control Panel** and hunting down the **Device Manager** icon. Once you've started Device Manager, click on the **Port** section and you'll see all your computer ports listed in order. Look for a port labeled "USB Serial Port" or "Virtual COM Port" (see **Figure 2.4**). Next to it you'll see a COM number. Write this number down because you'll need to enter it when you're setting up your JT65/JT9 program when it asks for the "serial port" or "PTT port" (**Figure 2.5**).

When the computer recognized the USB interface, it also recognizes the sound device within the interface. It will consider this device as another sound unit, just as though you had installed a second sound card inside the computer. (The computer doesn't know that this sound device is sitting in a box a few feet away and it doesn't care!) Again, this is important to understand because when you set up your software you may

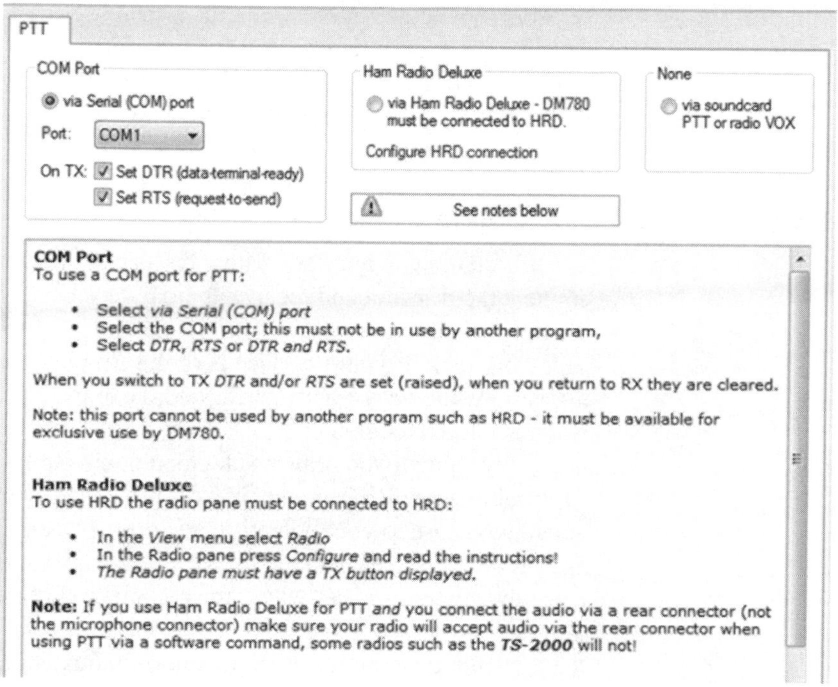

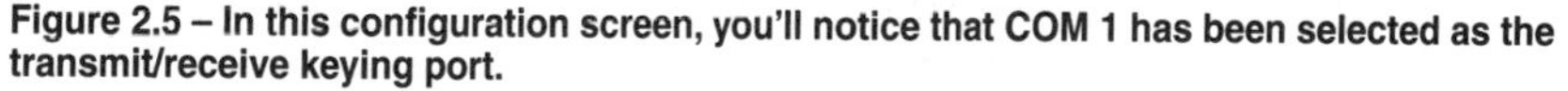

Figure 2.5 – In this configuration screen, you'll notice that COM 1 has been selected as the transmit/receive keying port.

need to specify which sound device the software should use. Obviously you will need to select the sound device in your interface. Most software applications have drop-down menus that will list the available sound devices automatically. Don't expect to see your interface device listed by brand name. Instead, it may show up as "USB Sound," "USB Audio Codec" or something similar.

Non-USB Interfaces

If your interface doesn't have a USB connection, it likely uses a traditional serial connection instead. If your computer has a serial (COM) port, you need only attach a serial cable (typically a cable with 9-pin plugs at both ends) between the computer and the interface.

In most computers these serial ports fall in a range between COM 1 and COM 4. Plug your serial cable into an available port and use a bit of trial and error to find out which one you've selected. Start your JT65/JT9 program and go to the configuration menu. Enter a "1" into the COM port selection box and then either run the transmit or PTT test (if the software provides this), or initiate a JT65 "CQ" to force the software to send a transmit command. Either way, if your transceiver goes into transmit, congratulations – you've found the correct COM port. If not, try 2, 3, 4, etc.

Managing the Audio Connections

If you own an interface with a built-in sound device, you'll need to connect the transmit and receive audio cables between the interface and the transceiver, either at the microphone and headphone jacks, or at the accessory jack. If your interface is of the simple transmit/receive switching variety, the one or both sets of audio cables may have to go all the way back to the computer.

Even though you're using shielded audio cables, there is the potential for trouble when RF is in the air. This is especially true when your station antenna is close to your operating position. The audio cables can act like antennas themselves, picking up RF and wreaking havoc on your station. I've seen some instances where the computer shut down or reset whenever the transceiver was keyed. In other examples the RF energy mixed with the transmit audio and resulted in a horrendously distorted output.

If you suspect you have an RF interference problem in your JT65/JT9 station, you can diagnose it by reducing your output power and observing the results. If 100 W output gives you grief but 50 W is smooth as silk, you clearly have an RF interference issue.

Presumably you've kept your audio cables are short as possible. Stringing up 20-foot-long audio cables between the radio, the interface and the computer is just asking for trouble.

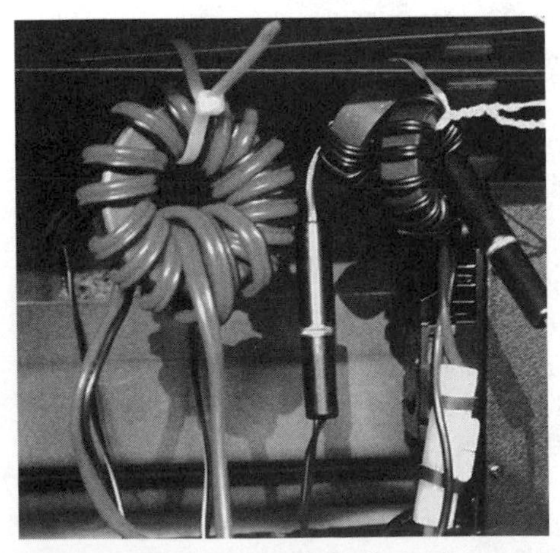

By wrapping cables through ferrite toroids, you can suppress or eliminate RF interference.

But if your audio cables are of reasonable length and you still suffer interference, it is time to buy some *toroid cores*. These are circular donuts made of a powdered iron and epoxy mixture. They come in various sizes and are rated for suppression at various frequencies. For HF applications, Type 61 toroids are among the most effective. To suppress RF on an audio cable, wrap the cable through the toroid at least 10 times with evenly spaced turns.

With right toroid in the right place, you can greatly reduce or eliminate RF interference. For severe cases you may need to place a toroid on every cable.

You'll often see used toroids for sale at hamfest fleamarkets, but don't buy a used toroid unless you know the type of material it contains. When in doubt, buy toroids new from manufacturers such as Amidon at **www.amidoncorp.com**. Avoid snap-on ferrite cores. While they are certainly easy to use, they are not as effective as toroids that you wind yourself.

Setting Up the Transceiver

Much of the advice that follows depends on what sort of transceiver you own and what kind of interface you are using to create your JT65/JT9 setup. When in doubt, always consult your transceiver and interface manuals.

If your interface is connecting to the radio through the transceiver accessory port, see if there is a function in the transceiver to adjust the accessory audio input and output levels. If it exists, this is a convenient way to establish "baseline" audio levels for the radio. Of course, you can also adjust audio levels at your computer and possibly at your interface (depending on the kind of interface you purchase). If it seems as though you aren't getting enough audio from the radio, or if it seems that you can't drive the radio to the desired output regardless of the computer or interface settings, check these transceiver settings as well.

Audio Overdrive

Speaking of driving rigs to full output, we need to discuss the danger of audio overdrive. Without question this is one of the most common issues among new JT65/JT9 operators.

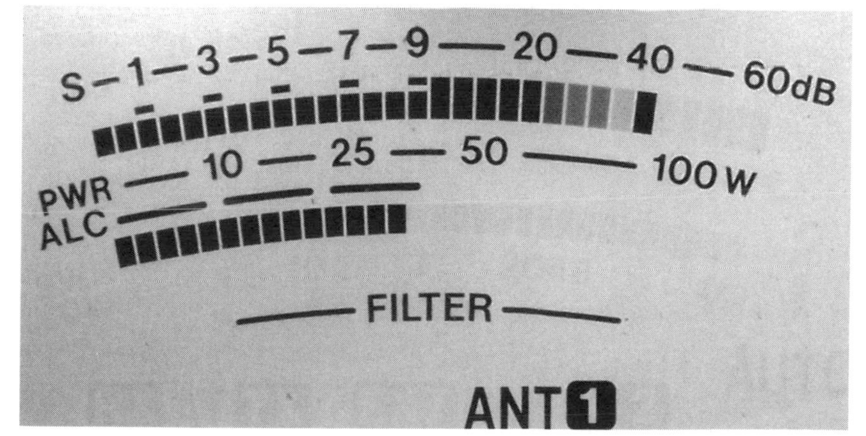

In the case of this Kenwood TS-2000 transceiver meter, the ALC metering scale is on the bottom. As you can see, we're overdriving the radio with excessive audio, causing a substantial amount of ALC activity. Ideally, this meter should be reading zero.

In this ICOM transceiver meter, the ALC "safe zone" is indicated with a red bracket. As long as the needle stays within the zone, you have a decent chance of transmitting a clean signal.

Rather than gazing at the transceiver meter as it displays your RF output, switch the meter to monitor *ALC* (Automatic Limiting Control) instead. All transceivers display ALC activity differently. The display may simply indicate the presence and amount of transmit audio limiting taking place. Other displays may include a "safe zone." If the ALC activity remains within the safe zone, your transmit audio levels are acceptable.

When you transmit, do not increase the audio level beyond the ALC safe zone, or beyond the point where ALC activity is excessive. When you see the needle or LEDs swing hard to the right, this is a warning that you are supplying way too much audio to the radio and that the ALC circuit is trying to rein you in.

The goal is to generate the desired RF output while keeping ALC activity to a minimum (or even zero), or while keeping the ALC meter in the safe zone. It is important to keep in mind that minimal ALC activity does *not* necessarily guarantee a clean signal. It does in many instances, but your best insurance is to ask for reports whenever you are in doubt.

In Chapter 4 we'll expand this discussion, especially as it concerns how your signals look and sound on the receiving end.

Computer Sounds

When you begin listening to JT65/JT9 signals, don't be surprised if you occasionally hear beep, dings, chimes or a disembodied voice declaring "You've got mail!"

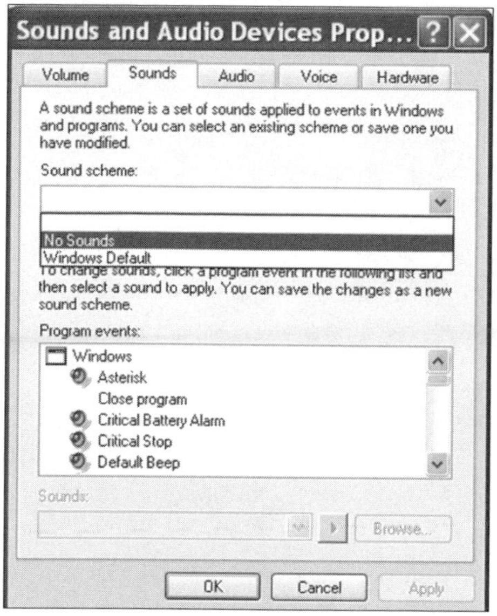

Most of this interference isn't deliberate. It is caused by VOX-type interfaces that key transceivers whenever they detect audio – *any audio* – from the computers. I'm willing to bet the operators aren't even aware that their computers are guilty of this obnoxious behavior.

The solution, at least in *Windows*, is simple: Turn off "*Windows* sounds" before you get on the air. You can also do this manually by opening *Window's* **Control Panel** and double clicking the **Sound and Audio Devices** icon. Click the Sound tab and under Sound Schemes select **No Sounds**. (Depending on the *Windows* version in question, the labels may differ.) Your fellow hams will thank you!

You can guard against transmitting odd *Windows* noises by turning off Windows sounds in Control Panel.

CAT Communication

If you want to allow your software to read your transceiver's frequency and perform other functions, you'll need to make sure that the data communication rates between the interface or computer and the transceiver are the same. Some clever pieces of CAT software will automatically analyze the data from the transceiver and quickly determine the data rate. For others, you'll have to enter a menu and specify the data rate.

In most CAT-capable radios you'll find a menu setting that will allow you to specify a data rate, often expressed as "baud." A rate of 9600 baud, for example, is common. You may need to access this menu to find out what setting the radio is currently using, or to change the radio's data rate to something the interface or software can handle.

Chapter 3

On the Air with JT65 and JT9

As we discussed in Chapter 2, most amateurs are using either Dr Taylor's *WSJT-X* software or Joe Large's *JT65-HF* application. Of course, if your interest is in working both JT65 *and* JT9, your only software choice at the moment is *WSJT-X*.

Software has a tendency to evolve over time. That makes it difficult to offer detailed instructions in a printed book without the risk of those instructions becoming obsolete soon after the book goes to press. Fortunately, *WSJT-X* and *JT65-HF* have a number of features in common, and I don't expect these to change dramatically in the near future. Also, Joe Large, W6CQZ, has suspended development of *JT65-HF*, so it is likely to remain stable for several years.

In this chapter I will concentrate primarily on the features of both programs that you are most likely to use on a regular basis. There are many other features that you can use to adjust the software behavior to your liking, but I'll leave those details to the user manuals.

The Importance of Time

As I mentioned earlier in this book, timing is extremely critical to JT65 and JT9. For these modes to work properly, your computer clock must be as accurate as reasonably possible. In both *JT65-HF* and *WSJT-X* you'll see displays that indicate how far out of sync your computer clock is compared to the clocks of other stations. When operating JT65, your computer clock cannot be more than two seconds out of sync with the clock at the station you wish to contact. Otherwise, nothing will decode. JT9 synchronization requirements are even tighter (no more than 1 second out of sync).

The good news is that while *Windows* computers in particular are somewhat sloppy timekeepers, they are accurate enough for JT65 and

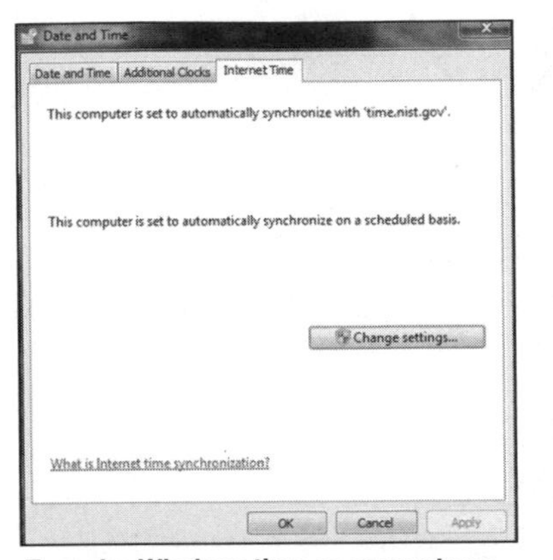

To make *Windows* time as accurate as possible, open Control Panel and click on Date and Time. In the Date and Time window, click on the Internet Time tab, and then click the Change Settings button.

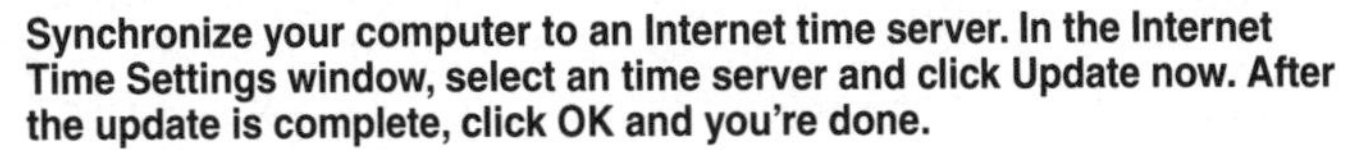

Synchronize your computer to an Internet time server. In the Internet Time Settings window, select an time server and click Update now. After the update is complete, click OK and you're done.

JT9. As long as the computer has a connection to the Internet, staying in sync is relatively easy. All you have to do is open *Window's* **Control Panel** and click on **Date and Time**. In the Date and Time window, click on the **Internet Time** tab, and then click the **Change Settings** button. In the next window select the Internet time server and click **Update now**. After the update is complete, click **OK** and you're done. Do this every time you sit down to operate and you'll always be in sync.

If you find this to be too much of a hassle, you can install a free program such as *Dimension 4*, which runs in the background and keeps your computer in sync with various time Internet services. You can download *Dimension 4* at **www.thinkman.com/dimension4/**.

Take a Moment to Configure Your Transceiver and Interface

Before you fire up your JT65/JT9 software, turn on your transceiver and make sure it is configured as follows:

• **RF Output Power: 50 W or less.** Not only do you not need more than 50 W to make contacts, keeping the output low will help protect your radio.

• **Speech processing OFF.** Never use speech processing with any digital mode.

• **Mode: USB** (Upper Sideband). JT65 and JT9 always operate with USB, regardless of frequency.

• **Interface: Plugged in and enabled.** Before you start your JT65/JT9 software, make sure your interface is plugged into your computer, including any transmit/receive audio cables you may be using. Depending on the type of computer you own, the software may need to "see" that the audio inputs and outputs are properly in place when it starts or it will generate a cryptic error message such as "Sound device not found or unavailable."

Finally, if you are using *JT65-HF* along with a piece of CAT software that reads the transceiver's frequency, start the transceiver software *before* you start *JT65-HF*. *WSJT-X* has its own CAT communication built-in, so this instruction only applies to *JT65-HF*.

First Time Set Up – *JT65-HF*

When you start *JT65-HF* for the first time, your initial task is to set up the program. Click **Setup** in the upper left corner. This will open the **Configuration** window. This window has seven tabs lined up horizontally, but we need only be concerned about a few of them.

The most important tab is **Station Setup**. Fill in your call sign and your grid square. If you don't know your grid square, you can find out online by using K2DSL's handy calculator at

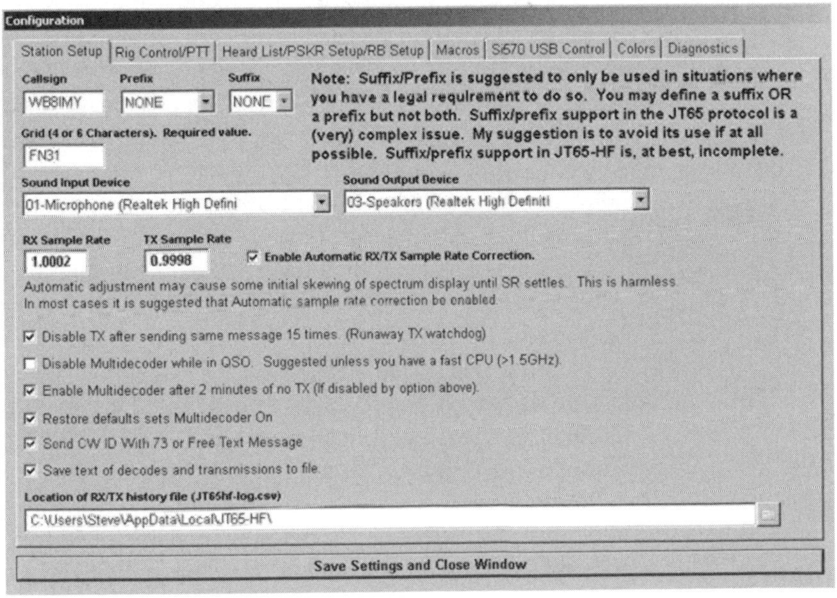

The *JT65-HF* station setup screen.

www.levinecentral.com/ham/grid_square.php. All you have to do is enter your ZIP code and the site will respond with your grid square.

Select your **Sound Input** and **Sound Output** devices. Click the drop-down arrows and you'll see a list of every device *Windows* recognizes. If you are using the sound card or sound chipset in your computer it will appear in this list, probably with a label such as "Microphone" or "Speakers." If you are using an interface with a built-in sound device, it may appear with an odd label such as "USB Audio Codec."

I suggest that you accept the default settings for all the checkboxes that appear below this section, particularly the one labeled **Enable Automatic RX/TX Sample Rate Correction**. This handy feature will automatically adjust your sound device sample rate when the program starts. You may see evidence of its activity in the first few seconds as lines in the waterfall display appear to skew to the left or right before settling down to a perfectly vertical orientation. You won't be able to decode signals during the first minute as a result.

Now click the **Rig Control/PTT** tab and fill in the COM port number for the line you are using to switch your transceiver between transmit and receive. You can test this function by clicking on the Test PTT button. When you do, your transceiver should quickly bounce into the transmit mode and then back to receive.

Below this section, you'll see a row labeled *Ham Radio Deluxe*, *OmniRig* and *Commander*. This is only important if you are going to have one of these three pieces of software running in the background to switch your transceiver between transmit and receive. In most instances you can ignore this row.

Next, click on the tab labeled **Heard List/PSKR Setup/RB Setup**. If your station has an Internet connection, *JT65-HF* will share information about the stations you are hearing so that other amateurs can observe propagation conditions, or test their equipment. Sharing this information is harmless and very helpful to the ham community at large. We will discuss this in more detail in the next chapter. For now, I'd suggest that you join the party by filling in your call sign and a short description of your antenna system.

On this page you'll also see yet another row labeled *Ham Radio Deluxe*, *OmniRig* and *Commander*. *JT65-HF* has the ability to read the operating frequency from your radio through one of these programs. You do ***not*** need this function to enjoy *JT65-HF*; it is more of a convenience feature, as you'll see later. If you are using one of these applications for rig control, go ahead and click "enable" in the appropriate box. Otherwise, leave it alone.

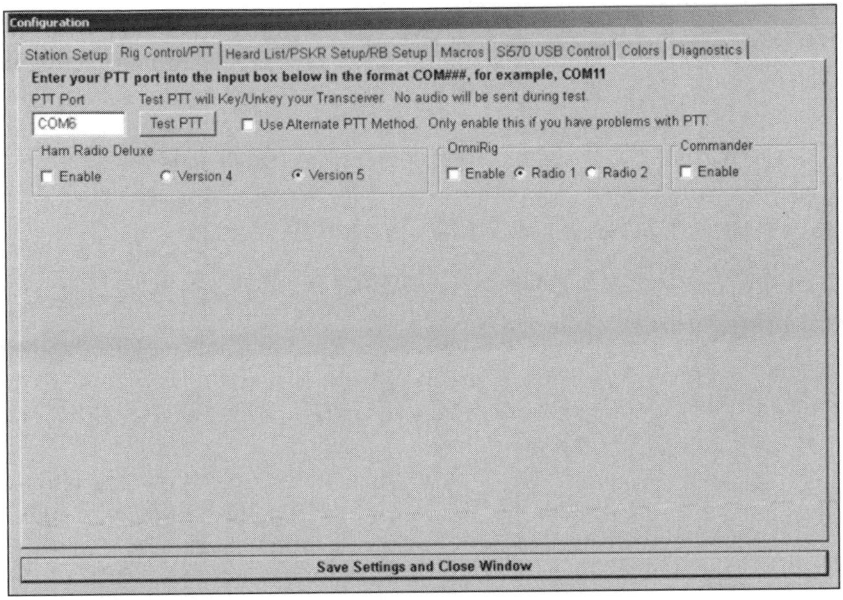

In the Rig Control/PTT tab of the *JT65-HF* setup screen, you can input the COM port you are using to key your radio and then test the functionality.

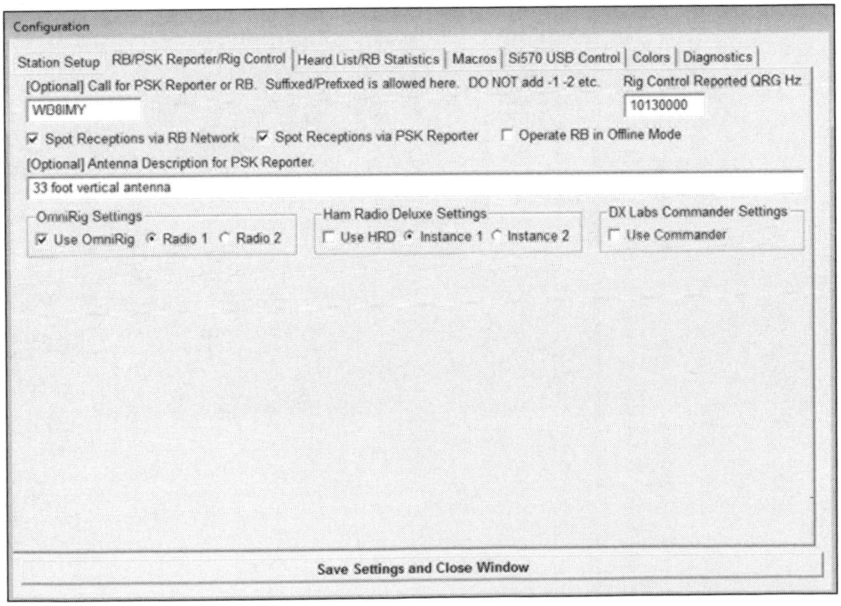

In this window you can configure *JT65-HF* to send signal reports to the Internet.

All the other tabs are safe to ignore at this time. Their functions allow you to customize *JT65-HF* and to keep things simple we're going to stick with all the default settings for now. Click **Save Settings and Close Window**.

A Tour of the *JT65-HF* Main Screen

Let's take a look at the *JT65-HF* main screen, section by section, starting at the top. See **Figure 3.1**.

The waterfall display dominates most of the top portion of the *JT65-HF* main screen. Whenever *JT65-HF* is running, it sweeps through your receive audio spectrum from 0 to 2000 Hz. Every signal it detects appears in this window.

You'll notice that the waterfall is divided into two halves to the right and left of the center "zero" point. The display markers are positive to the right of the zero (0 to 1000) and negative to the left of the zero (0 to -1000). Along the top of the waterfall you'll see a red and green bracket. If you click your mouse cursor within the waterfall the bracket will move to the position you just clicked. The bracket represents your transmit/receive window.

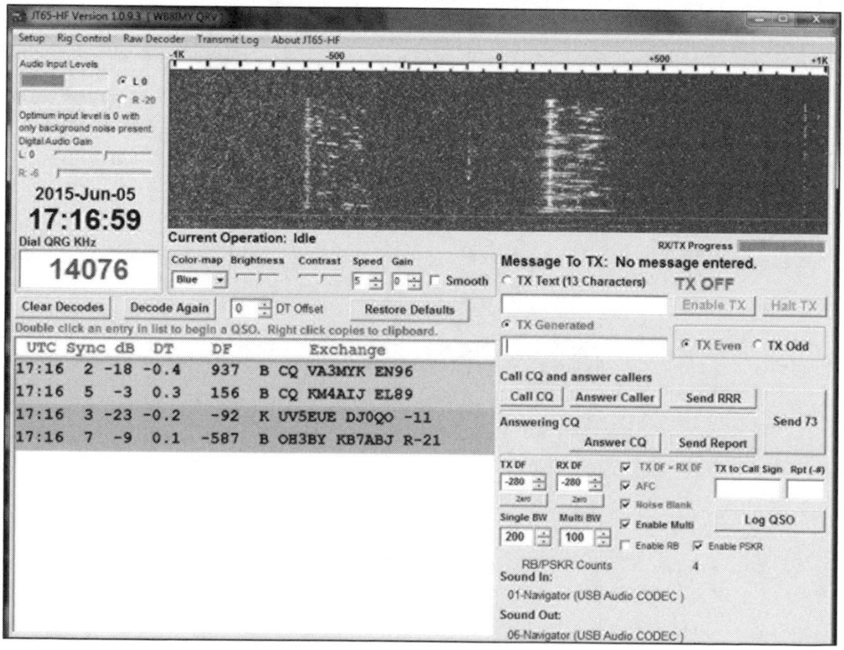

Figure 3.1 – The main *JT65-HF* screen.

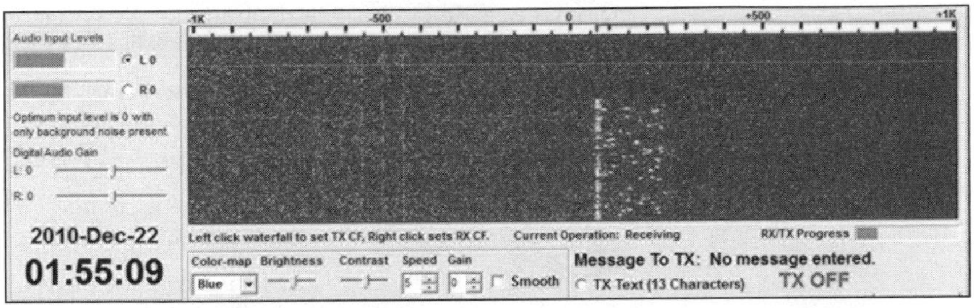

Figure 3.2 – On the left side of the waterfall you have the right and left channel audio input controls.

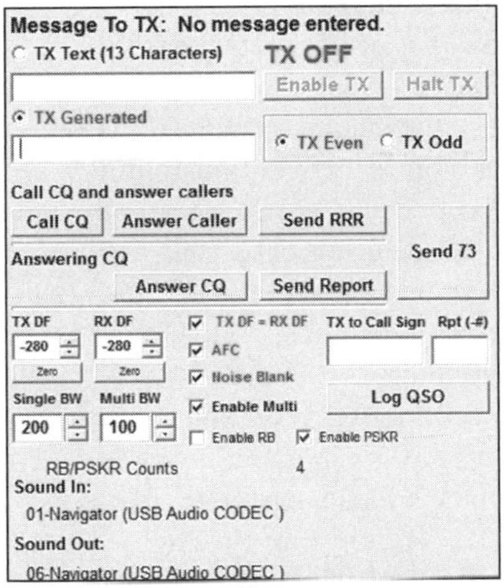

Figure 3.3 – In the lower right section of the *JT65-HF* screen you'll find a number of check boxes and buttons (see text).

JT65-HF can operate in simplex (transmitting and receiving on the same frequency) and split (transmitting and receiving on different frequencies). Most of your contacts will be simplex, but it is worthwhilc to know that *JT65-HF* has split-frequency capability. Two brackets appear when operating split – red for the transmit frequency and green for the receive frequency.

On the left side of the waterfall (**Figure 3.2**), you have the right and left channel audio input controls. When the band is quiet (when there are no signals), you should adjust the right and/or left channel controls to achieve 0 dB on both channels. Note that some interfaces supply audio on either the right or left channels, but not both. This is fine; just make your level adjustment to the active channel and set the inactive channel to zero. Keep in mind that too much or two little audio makes it difficult to decode signals. As Goldilocks observed in the children's tale, the best porridge was the one that was not too hot and not too gold – just right.

Just below the audio level controls you'll see the Date, Time and the Dial QRG displays. If you have set up *JT65-HF* to communicate with your transceiver, it will fill in the operating frequency automatically. If not, you should manually enter your frequency here.

Below the waterfall you'll see controls for color, brightness, contrast, speed and gain. You're safe leaving all these at their default settings. There is little need to change them unless you're having difficulty viewing the waterfall, or unless you are running *JT65-HF* on a slow computer.

Moving to the lower right section (**Figure 3.3**) we have a number of check boxes and buttons. These may seem confusing, but their functions will become more apparent when we make our first contacts.

TX Generated is the text you are sending to the other station. This window is for transmitted text that *JT65-HF* generates automatically, either when you click on one of the buttons below, or when you click on a line in the decoding window. Again, the function of this window will become clear as we step through your first contact. Just above this window you'll notice the **TX Text** window. This is for manual, non-automatic text, otherwise known as *free hand* text.

To the right of the message windows are buttons to enable or halt transmission. Below these buttons is the section that allows you to choose whether you wish to transmit on even or odd minutes. With JT65 you transmit and receive in turns – during one minute you transmit and during another minute you receive. Therefore, one minute may be an odd-numbered minute (such as 2105 UTC) and the next minute may be an even-numbered minute (such as 2102 UTC).

In most instances *JT65-HF* will make the choice for you. However, when calling CQ you get to choose when you will begin transmitting – on an even or odd minute. You definitely don't need to worry about even and odd minutes when you are answering someone else's call. That station has already selected which minute (even or odd) he will use. When you double click your mouse on the red or green line to respond, *JT65-HF* will automatically choose the opposite minute.

Below the text-generating buttons you'll find several other interesting sections. The **TX DF** and **RX DF** sections are for split-frequency applications. For the vast majority of your JT65 contacts you will work simplex, so you want to leave the **TX DF = RX DF** box checked.

Below the TX DF and RX DF sections you'll find the controls for **Single Decoder BW** (Bandwidth) and **Multi Decoder Spacing**. This is another set of controls you'll rarely have a reason to change from their default values. The default of 200 Hz for single bandwidth is usually adequate.

Put a checkmark in the **AFC** (Automatic Frequency Control) box so that *JT65-HF* can compensate for stations that may be drifting a bit. If you live in an area with frequent storms or other sources of noise, put a checkmark in the **Noise Blank** box as well.

Enable Multi refers to *JT65-HF*'s ability to decode many signals at once. When enabled, the decoder will attempt to decode all JT65 signals within the 2 kHz passband. Unless you are using a slow computer that has difficulty processing so much information at once, always leave this box checked.

To the right you'll find the Log QSO button. When you click this button *JT65-HF* will save your contact information (who you worked, when, etc) to the file **jt65hf_log.adi** in the *JT65-HF* directory in a

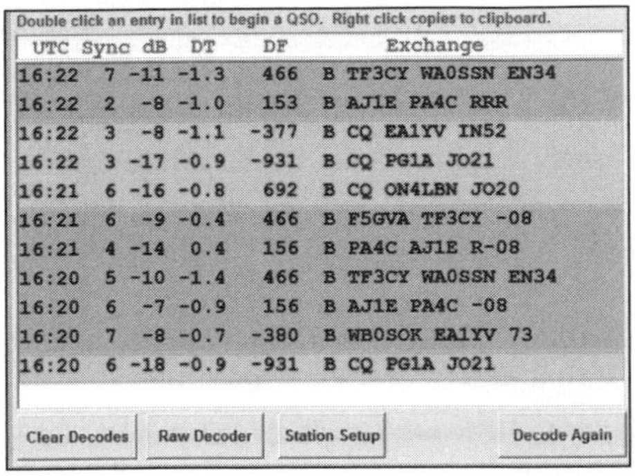

Double click an entry in list to begin a QSO. Right click copies to clipboard.

UTC	Sync	dB	DT	DF		Exchange
16:22	7	-11	-1.3	466	B	TF3CY WA0SSN EN34
16:22	2	-8	-1.0	153	B	AJ1E PA4C RRR
16:22	3	-8	-1.1	-377	B	CQ EA1YV IN52
16:22	3	-17	-0.9	-931	B	CQ PG1A JO21
16:21	6	-16	-0.8	692	B	CQ ON4LBN JO20
16:21	6	-9	-0.4	466	B	F5GVA TF3CY -08
16:21	4	-14	0.4	156	B	PA4C AJ1E R-08
16:20	5	-10	-1.4	466	B	TF3CY WA0SSN EN34
16:20	6	-7	-0.9	156	B	AJ1E PA4C -08
16:20	7	-8	-0.7	-380	B	WB0SOK EA1YV 73
16:20	6	-18	-0.9	-931	B	CQ PG1A JO21

Clear Decodes | Raw Decoder | Station Setup | Decode Again

Figure 3.4 – The *JT65-HF* signal decoding window.

standard ADIF format that you can import into your computer logging software.

At the bottom of the lower right portion of the *JT65-HF* window you'll see a line of text that reads **RB/PSKR counts** and two checkboxes labeled **Enable RB** and **Enable PSKR.** If you've checked the **Enable RB** and **Enable PSKR** windows *JT65-HF* will automatically access your home Internet connection and share your data (the stations you've heard and how strong they were) with W6CQZ's reverse beacon (RB) website – if it is active – and with the PSK Reporter (PSKR) website – *if* you've enabled this feature in the station setup screen. The information is extremely helpful to your fellow amateurs who study propagation, experiment with new antennas and so on.

Finally, look at the decoding window in the lower left corner (**Figure 3.4**). This is probably the most important part of *JT65-HF* because this window will display the information about the signals you are receiving, including any message texts. Notice the labels along the top of the window…

UTC: The time the signal was decoded (in UTC).

Sync: This is a measurement of the strength of the synchronizing tone. The higher the number, the better the sync signal.

dB: The strength of the JT65 signal in decibels. The lower the number, the stronger the signal. Zero dB is the strongest possible signal.

DT: How much the decoded station's time deviated from your time, measured in seconds or fractions of seconds. Ideally the decoded stations should be within 2 seconds of your computer's time, preferably less than 1 second.

DF: How far the signal frequency deviates, in Hertz, above or below the zero center point of the waterfall display. A negative number is a signal to the left of zero; a positive number is a signal to the right of zero.

Exchange: The information the transmission contained. If you see two call signs, the transmitting station is the ***second*** call sign.

Just to the left of the exchange text you'll see either a **B** or a **K**. This is a reference to the kind of error correction algorithm that *JT65-HF* used to validate the text. B stands for *BM*, a simple Reed Solomon algorithm. K means *KVASD*, a much more complex algorithm. One way to think about this is to imagine that a K means that *JT65-HF* had to work particularly hard to make sense of the signal. If so, this station may present a challenge if you attempt to complete a contact.

First Time Set Up – *WSJT-X*

When you start *WSJT-X*, you will see two windows appear more-or-less simultaneously. One window holds the waterfall display in which you'll see the JT65 and JT9 signals. The second window, which usually appears below the waterfall, displays all the program controls as well as the decoded signal information.

First, click **Files** in the upper left corner of the lower window, and then **Setup**. You will be presented with a new window containing six tabs. Open the **General** tab first (**Figure 3.5**). This is where you will fill in your call sign and grid square. For now, leave the other options at their default settings. If you don't know your grid square, you can determine it at **www.levinecentral.com/ham/grid_square.php**.

Next, click the **Radio** tab (**Figure 3.6**). The **Radio** window allows you to enter the settings for CAT control of your transceiver (on the

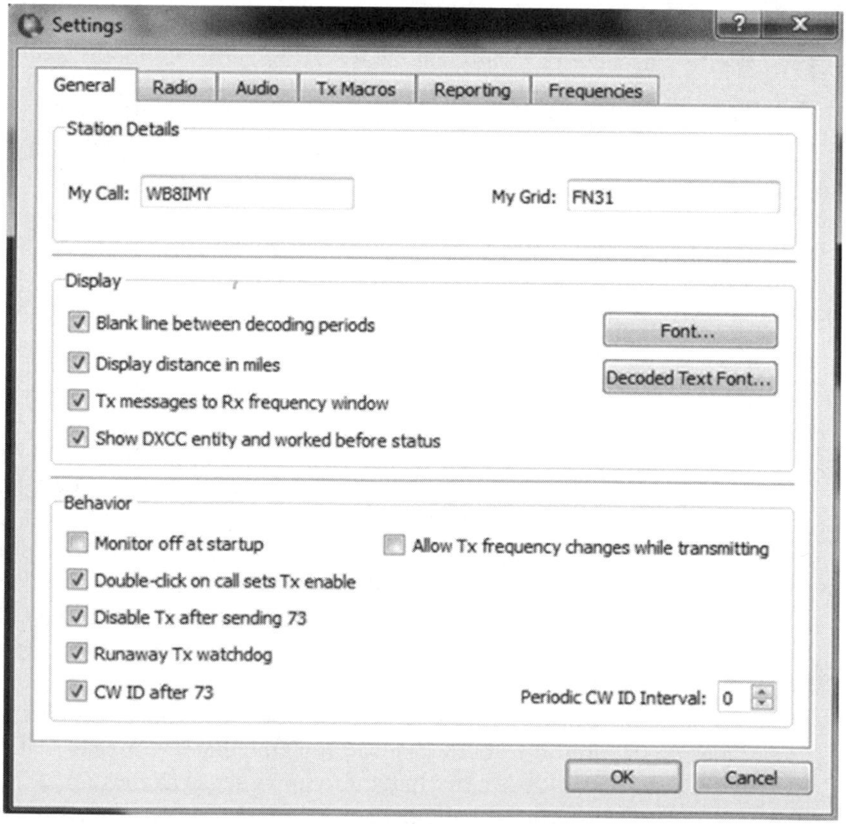

Figure 3.5 – The General section of the *WSJT-X* setup window.

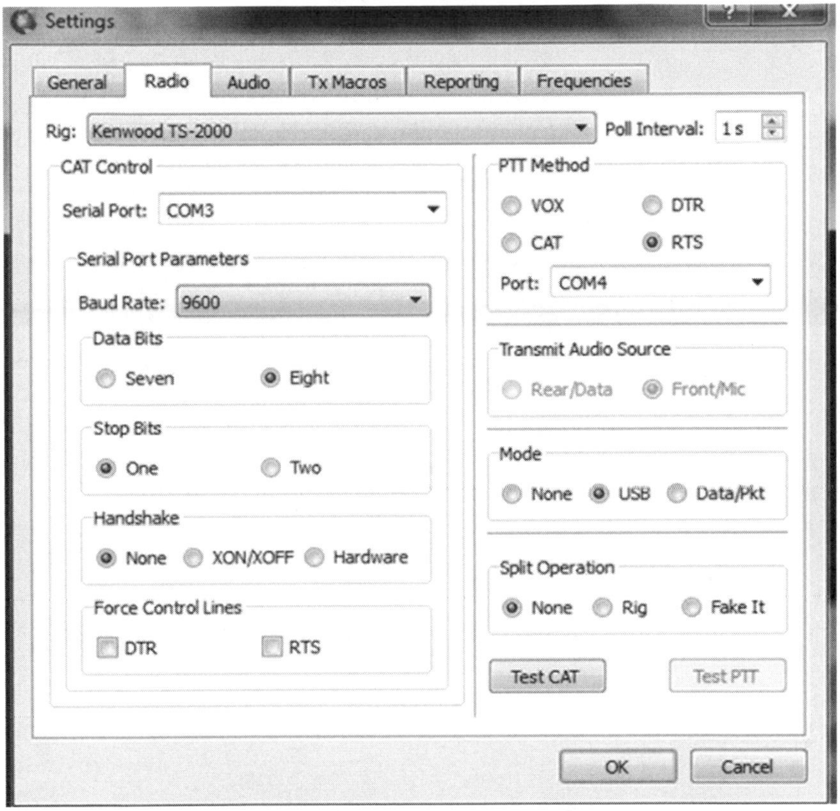

Figure 3.6 – The Radio window allows you to enter the settings for CAT control of your transceiver and the settings for PTT control.

left side of this window) and the settings for PTT (Push to Talk) control (on the right side of this window). As with *JT65-HF*, CAT control is *not* mandatory, although it is awfully convenient.

In the example shown, I have set up *WSJT-X* for CAT control of my transceiver using a 9600 baud signaling rate on COM 3 with eight data bits, one stop bit and no handshake parity. You'll find that this is a common setting with only the COM port and the signaling rates being the variables. In my case, my interface supports CAT control on COM 3 and my transceiver communicates at 9600 baud. Your settings will likely be different.

On the right side of the **Radio** window, under "PTT Method," I have chosen COM 4 as my Push to Talk control port. Once again, your settings will likely be different. You may have that you have to do some experimentation to find the correct values for both CAT and PTT.

The **Audio** window (**Figure 3.7**) allows you to select your audio input and output devices. In this example, I am using a Timewave Navigator interface, which has a built-in sound device. Therefore, it appears in the list of options when I click the drop-down menus. You can also select right channel, left channel, mono or "both" audio streams if you are using an interface that sends audio in a particular fashion. Once again, this may require a little experimentation.

Let's skip **TX Macros** and go to the **Reporting** tab (**Figure 3.8**). This is where you can tell *WSJT-X* to automatically send activity reports to Internet reporting sites such as PSKReporter, which we'll discuss in the next chapter. If you have Internet connectivity at your station, it is always a good idea to activate this feature. The more operators that share data, the more useful these sites are for everyone. With that in mind, make sure the "db reports to comments" box is checked. This adds your signal report data to the reports you send to the Internet.

We won't bother with the **Frequencies** tab, so at this point we are done. Just click **OK** to save all the settings.

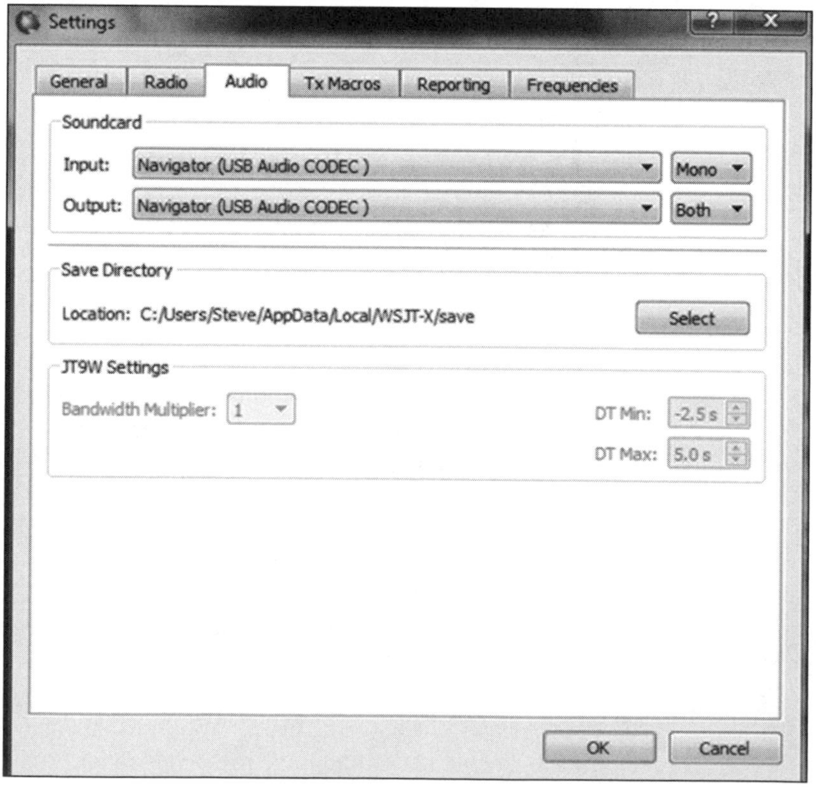

Figure 3.7 – The Audio window allows you to select your audio input and output devices.

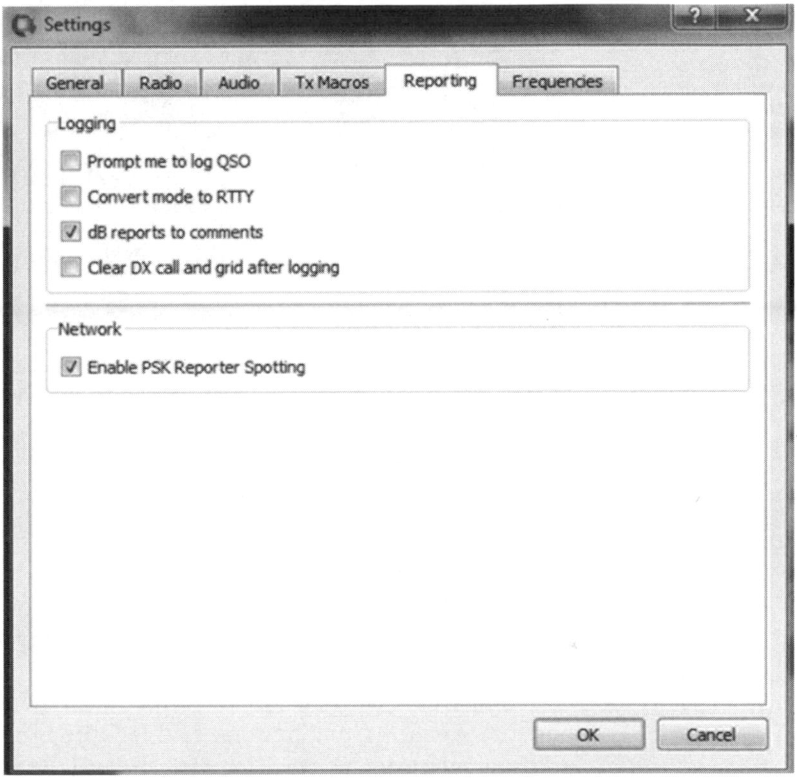

Figure 3.8 – The Reporting tab presents a screen where you can configure *WSJT-X* to automatically send activity reports to Internet reporting sites such as PSKReporter.

A Tour of the *WSJT-X* Main Screen

As it turns out, *JT65-HF* and *WSJT-X* have much in common when it comes to transmitting and receiving. If you compare the *JT65-HF* screen images in this chapter to *WSJT-X*, you will see many features that are familiar.

However, one of the first things you may notice is that unlike *JT65-HF*, the *WSJT-X* waterfall appears separately from what we'll call the *primary window*. The waterfall is also not divided into 1000 Hz sections to the right and left of a center point. Instead, the *WSJT-X* waterfall begins at zero Hz on the left and goes to about 3000 Hz on the right in one continuous sweep. As a result, *WSJT-X* doesn't use a DF value to indicate the position of the signal; it simply displays the signal frequency in Hz. Finally, the waterfall window has its own set of controls. You can leave these are their default values. Most *WSJT-X* users never bother to touch them.

When *WSJT-X* starts you are presented with two windows: The primary at the bottom and the waterfall at the top. You can reposition both windows as you prefer.

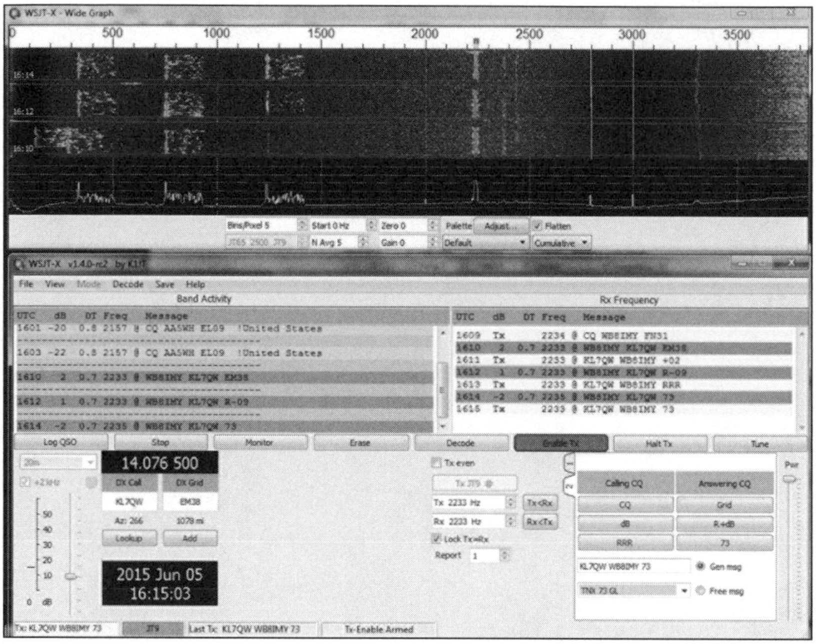

Now let's look at the primary window. In the upper left corner you'll find the menus. Starting from the left, the **File** menu allows access to the **Settings** submenu that we discussed earlier. It also allows you to load and save files.

To the right you'll see the **View** menu. This menu allows you to display an "astronomical data" sub-window, but this is only of interest for moonbounce work.

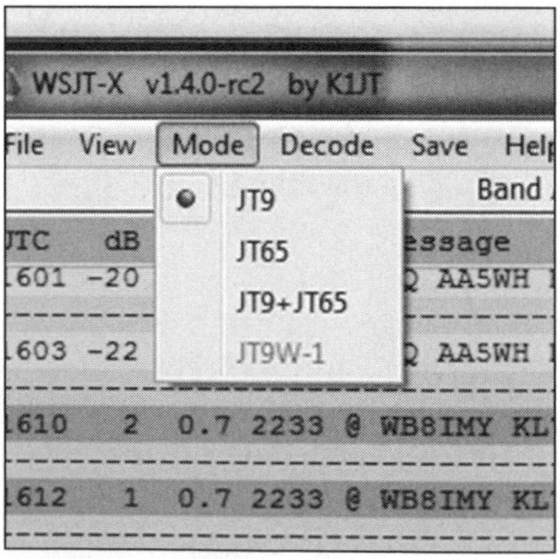

Look to the right of the **View** menu and you'll spot the **Mode** menu (**Figure 3.9**). In this menu you select your operating mode: JT65, JT9, JT65 and JT9 simultaneously, or any experimental modes that may be included (depending on the *WSJT-X* version you've installed).

To the right of the **View** menu is the **Decode** menu (**Figure 3.10**). As you will soon learn, the ability to decode signals quickly can be critical. You will have, at most, 12 seconds to decode the received signals, examine at the results, and then decide what to do next. If the band is active with many signals in the waterfall, *WSJT-X* is going to put substantial demands on your computer as it attempts to decode all the information. In this environment, an older, underpowered PC won't

Figure 3.9 – In the Mode menu you can select your operating mode: JT65, JT9, JT65 and JT9 simultaneously, or any experimental modes that may be included.

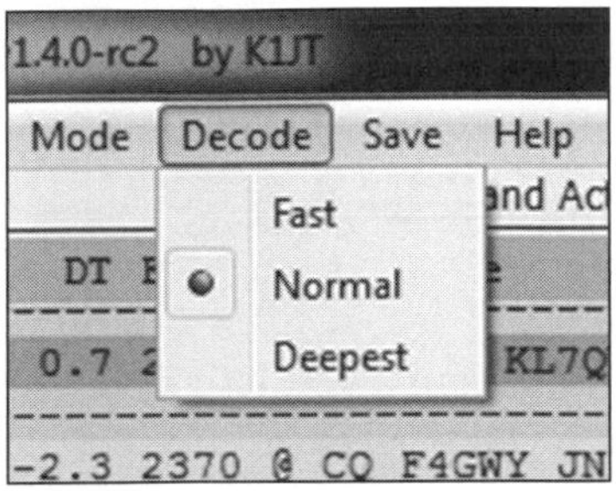

Figure 3.10 – The *WSJT-X* Decode menu.

be able to decode quickly and you'll lose precious seconds as you wait to see the results.

Generally speaking, if you are using a slower computer, choose **Fast**. Yes, you'll miss some signals, but at least you will have time to act upon those you can see. **Normal** will work well for most modern machines. I'd only recommend **Deepest** if you have a powerful computer. I have a capable station computer, but I find that if I choose **Deepest** *WSJT-X* will still be decoding when the next minute begins. That's far too long.

The **Save** menu offers some clever functions. You can use **Save all** to save all your received data as audio (WAV) files for loading and playback/decoding later. Or, you can select **Save decoded** to save only the data that contains at least one decoded message.

The **Help** menu is self explanatory.

Below the menu bar you will see the **Band Activity** and **RX Frequency** windows. These are the *WSJT-X* windows you'll be watching most often.

Similar to the *JT65-HF* software, the Band Activity window (**Figure 3.11**) includes a bar with the following labels (from left to right):

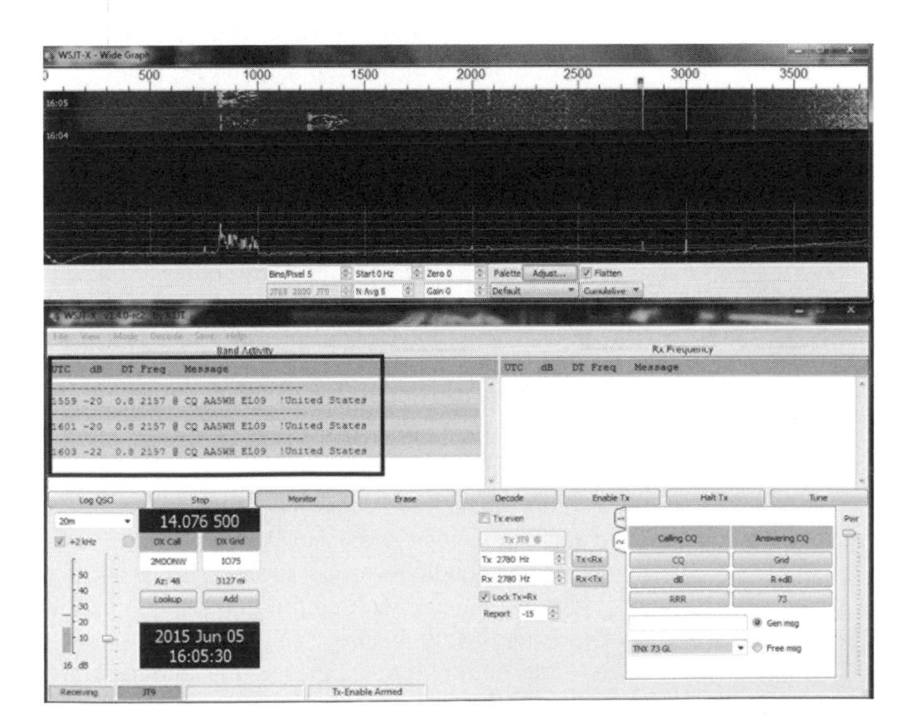

Figure 3.11 – The *WSJT-X* Band Activity window (highlighted within the black square). You'll be watching this window often!

UTC: The time the signal was decoded in UTC.

dB: The strength of the JT65 or JT9 signals in decibels. The lower the number, the stronger the signal. Unlike *JT65-HF*, *WSJT-X* will display signals that are stronger than 0 dB.

DT: How much the decoded station's time deviated from your time, measured in seconds or fractions of seconds.

Frequency: The position of the signal (in Hz) in the waterfall.

Message: The information the transmission contained. At the left edge of this field you have the mode flag. JT65 signals are market with pound signs (#) and JT9 signals are flagged with ampersands (@). This is followed by message text, call signs, signal reports and more. Remember: If you see two call signs, the transmitting station is the ***second*** call sign.

Signals shown in the Band Activity window with green backgrounds are CQ transmissions. Signals with red backgrounds are transmissions sent to *you*. (The use of red is intended to get your attention!) All other signals default to gray. Of course, you can change these colors if you wish.

WSJT-X has another neat feature that appears in the Band Activity window. As you may recall, *WSJT-X* and *JT65-HF* both have the ability to log contacts. *WSJT-X*, however, will take the additional step of searching the log and displaying interesting information in the form of enhanced colors or symbols at the far right-hand side of those CQ texts:

! (and a bright green background) means that this is a new DXCC entity for you.

~ (and a medium green background) means that you have already worked this DXCC entity, but not this particular station.

A dull green line without a symbol means that you have previously worked this station.

The RX Frequency window is not so verbose. It displays your transmissions (with yellow backgrounds), transmissions sent to you (with red backgrounds) and any other signals that happen to appear in the waterfall where you've positioned the red and green receive/transmit window bracket. We will discuss the use of this bracket later.

Below these windows you'll find a row of buttons. In normal operation you may not use these buttons very often, with the possible exception of Log QSO, which saves the contact information in the *WSJT-X* log and Tune, which enables a test transmission so that you can make sure your transmit audio levels are properly adjusted.

I'd recommend setting up *WSJT-X* to jump to the Monitor mode whenever you start the program. Should you forget, however, just click the Monitor button. If you are using the automatically generated messages to step through your contacts (we'll describe this in detail later), you won't need to click the Enable TX button to begin transmitting. If you are sending free hand text, however, you may need to use this button to enable transmission manually.

In the lower left portion of the primary window you'll find a vertical "slider" control. At the bottom of the control you will see a number indicating background noise level in dB. Click on the slider and drag it to the midpoint. Now, *when there are no signals present*, adjust your receive audio level at your interface (if it has such a control), the RF gain at your radio, or the receive audio level at your computer until the background noise level reaches 30 dB on vertical audio meter to the left of the slider. If necessary, you can use the slider to touch up this adjustment.

To the right of the slider there are displays for time, date, frequency and the call sign of the station you are contacting. *WSJT-X* will also attempt to compute how far away the station is and provide an azimuth beam heading in case you are using a directional antenna.

In the opposite corner of the primary window you will see the section that allows you to send either automatically generated or free-hand text (**Figure 3.12**). I prefer to use the tab marked "**2**." These are the messages you will normally use. Again, if you are using the automatic message

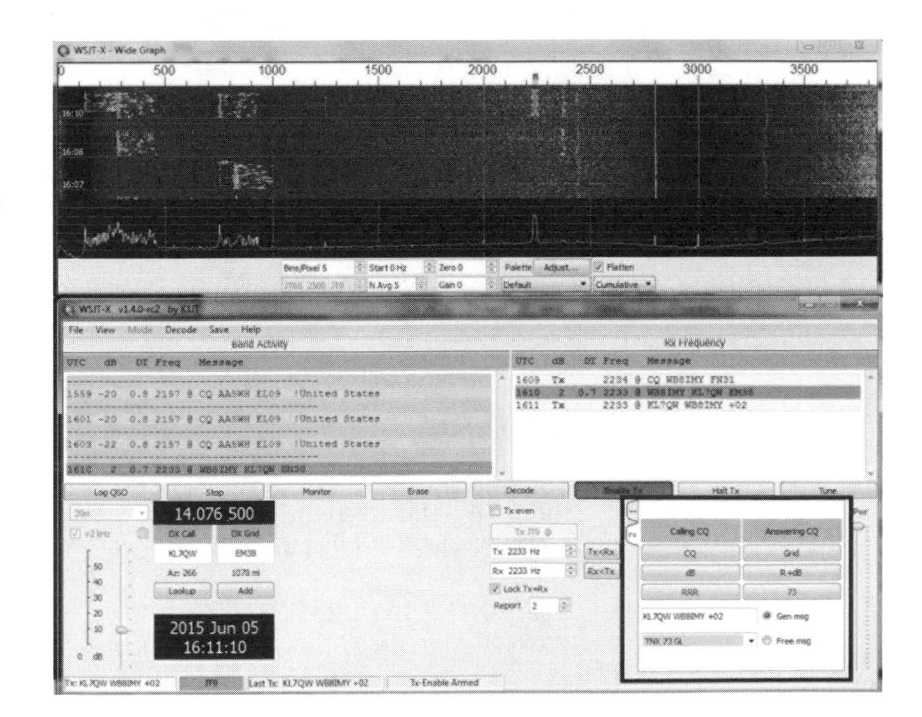

Figure 3.12 – The area highlighted in this image with a black border is where you will select automatically generated messages in *WSJT-X*, or send free-hand text.

generation feature, you won't have to touch these buttons, but they are awfully handy if things become "non-routine."

Finally, *WSJT-X* includes a convenient slider along the far right side of the primary window. Labeled "Pwr," this slider adjusts your transmit audio level. If your interface doesn't have a transmit audio control, the Pwr slider is an excellent substitute.

The JT65/JT9 "Dance"

Regardless of the software you are using, JT65 and JT9 contacts usually follow a strict sequence, not unlike a choreographed dance routine.

Let's try a hypothetical example. The dance begins with the first transmission at 2102 UTC. Keep in mind that when two call signs appear, the call sign of the transmitting station is the one on the **right**…

2102 CQ WB8IMY FN31

WB8IMY has begun sending CQ on an even minute from grid square FN31.

2103 WB8IMY N1NAS EN72

N1NAS replies and tells WB8IMY that he is located in grid square EN72.

2104 N1NAS WB8IMY -11

WB8IMY replies with a signal report of -11 dB.

2105 WB8IMY N1NAS R-15

N1NAS acknowledges the signal report from WB8IMY with an "R" followed by a signal report (-15).

2106 N1NAS WB8IMY RRR

WB8IMY sends "RRR," which means "Roger, roger, roger." Everything has been received and the exchange is complete.

2107 WB8IMY N1NAS 73

N1NAS sends 73 – best wishes.

2108 N1NAS WB8IMY 73

WB8IMY sends his 73 as well. The contact has ended.

When both operators follow this script, completing a contact with either *JT65-HF* or *WSJT-X* is a matter of simply double-clicking your mouse cursor on the red background text as soon as it appears. (Remember: You have less than 12 seconds to act!) The software will generate the text for the next transmission based on what it "sees" on the line that you have selected and will enable transmission at the start of the next minute.

However, some operators don't follow this script line-by-line. I call these "non-routine" contacts. For example, some amateurs skip the RRR exchange completely. Once they have successfully exchanged signal reports, they go directly to the 73 message.

You may also see something unexpected such as "WB8IMY RRR 73." Double clicking on one of these lines may not generate the response you expect from the software. This is when the text buttons in *WSJT-X* and *JT65-HF* can be lifesavers. If all else fails, I usually do a quick click on the 73 button.

Text such as **<CALL SIGN> RRR 73** usually appears because the operator has revised his standard transmit macros (see the example in **Figure 3.13**). You can do this as well, but be careful not to create something that is likely to baffle the operator on the other end.

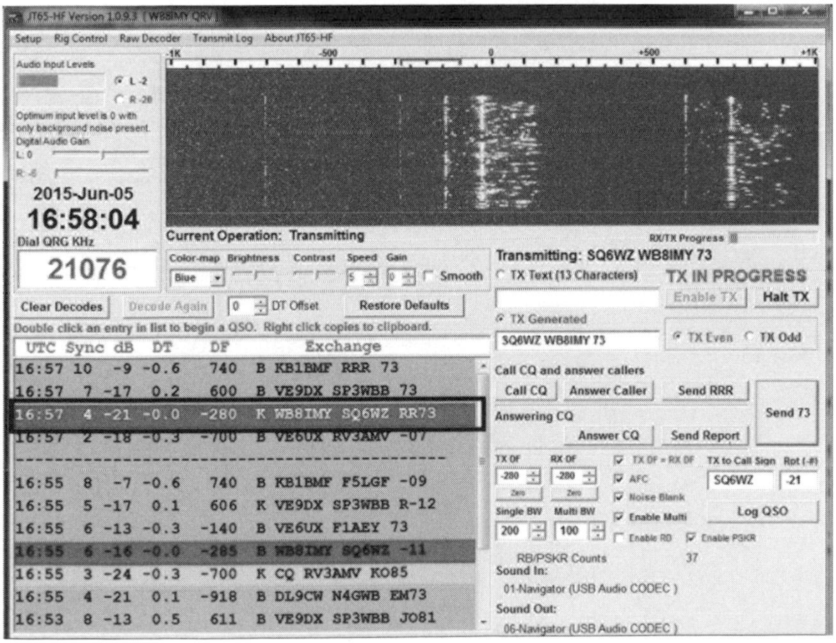

Figure 3.13 – In the text highlighted within the black square, you can see that SQ6WZ chose to skip the RRR part of the sequence and simply sent RR73.

You'll also see stations sending bits of free-hand text (see **Figure 3.14**). They are often doing this by typing the text into the **TX Text** or **Free msg** windows and then selecting this text to be sent instead of an automatically generated message. You may see something like **40W LOOP ANT**, which is shorthand for "I'm running 40 W to a loop antenna."

Notice that there is no call sign included in the free text example. So how do you know who actually sent it? Was it really intended for you?

If you are running *WSJT-X*, look for the frequency of a transmitting station that matches the frequency of the message. It may not be an exact match, but it will be close. Chances are, that's the station that sent the free-hand text. In *JT65-HF*, match the DF value of the free-hand text transmission to a recent transmission that includes a call sign (**Figure 3.15**).

Strange as it may sound, it is possible to conduct a conversation – sort of – using only free-hand text (see the example in **Figure 3.16**). One time I had a fellow ask about my antenna (I saw **UR ANTENNA?**). I replied with **DELTA LOOP** and we went back and forth for a bit – all with questions and answers of 13 characters or less (including spaces). It was tedious and nothing I'd recommend, but it *can* be done.

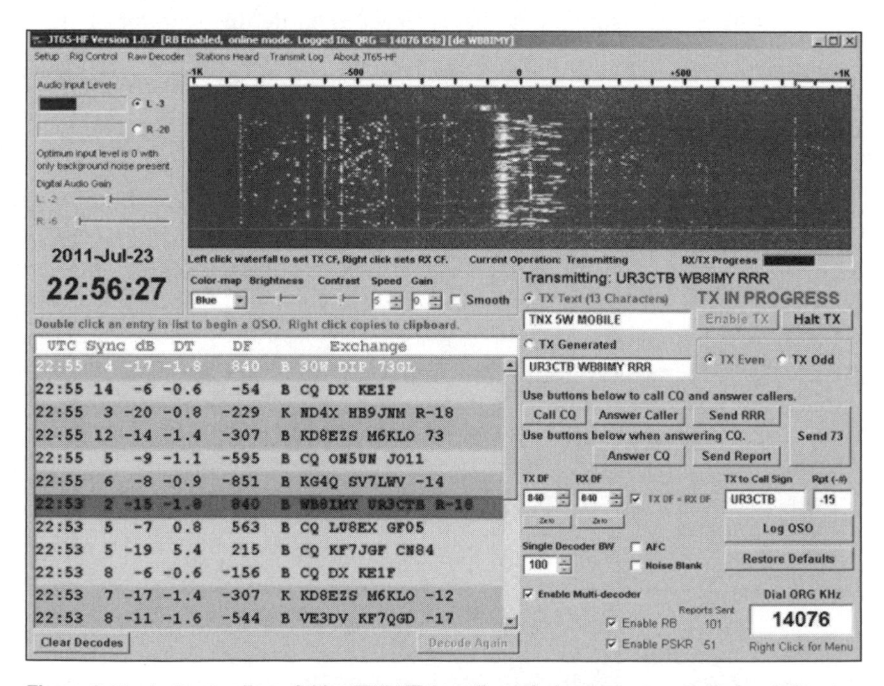

Figure 3.14 – In the top line of this *JT65-HF* decoding window you can see that a station has sent a line of free-hand text: 30W DIP 73GL. This translates to "I'm running 30 W to a dipole antenna. 73 and Good Luck."

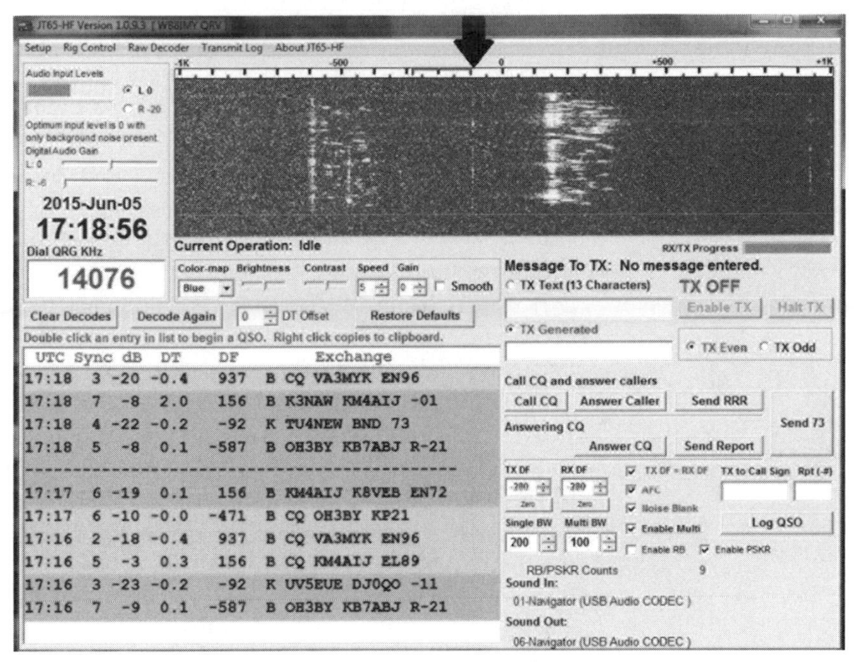

Figure 3.15 – Look in the *JT65-HF* decoding window and find the third line down from the top, which was sent at 17:18 – an *even* minute. You'll see that someone sent TU4NEW BND 73 (Thank you for the new band. 73). Notice that the DF is -92, or 92 Hz to the *left* of the zero center point. In this image, the signal is marked with a black arrow. Now glance down the decoding window at the *previous* even minute, which was 17:16. You'll see that DJ0QO was transmitting at -92 Hz. Chances are, the free-hand text at 17:18 was sent by DJ0QO.

JT65-HF Version 1.0.9.3 | W8BMY QRV

2015-Jun-07

16:24:57

Dial QRG KHz

50276.5

UTC	Sync	dB	DT	DF	Exchange
16:20	9	-19	-0.3	323	B CQ WB2YDS FN30
16:18	14	-16	-0.3	323	B CQ WB2YDS FN30
16:16	8	-19	-0.3	323	B CQ WB2YDS FN30
16:14	3	-18	-0.3	320	B CQ WB2YDS FN30
16:12	5	-19	-0.3	320	B OK 73 THEN
16:11	18	-4	-0.3	326	B NOT TODAY SRI
16:10	12	-18	-0.3	320	B 73 146.52 ?
16:09	19	-4	-0.3	326	B 5BAND 73 TOM
16:08	9	-16	-0.3	320	B AA6E WB2YDS RRR
16:07	22	-9	-0.3	824	B WB2YDS AA6E R-01

Figure 3.16 – In this example, we have two local stations conducting a free-hand text conversation on 6 meters between 16:09 and 16:12. One station was asking the other if he could switch to 146.52 MHz for a voice chat.

Eavesdropping on JT65 and JT9

As with many other Amateur Radio activities, the best way to become acquainted is to listen. I'd strongly recommend that you take your time getting familiar with your chosen software by using it to monitor activity.

You will find a list of popular JT65 and JT9 frequencies in **Table 3.1**. You might want to try 14.076 MHz as your initial destination since it is often the most active spot.

If you are using *WSJT-X* software, you have the ability to decode JT65 and JT9 signals simultaneously. The ability to do this is dictated by your receive audio bandwidth.

Most JT9 signals are usually 2 kHz above the JT65 "watering holes." For instance, with JT65 operators gathering around 14.076 MHz, you'll find the JT9 guys at 14.078 MHz. Assuming that your radio has about a 2.4 kHz receive audio bandwidth in the USB mode and you park the VFO at 14.076 MHz, you will receive signals from 14.076 MHz all the way up to almost 14.078.4 MHz – more than enough room to capture JT65 *and* JT9 transmissions.

Regardless of the software you are using, or which mode you've chosen, just dial up the correct frequency, adjust your receive audio level (wait until no one is transmitting) and then sit back and enjoy the show. The waterfall will slowly advance downward as the seconds tick by. JT65 signals will appear as solid lines with dots to the immediate right; JT9 signals show up as solid, wavering lines.

At the beginning of each new minute, the software will create a red horizontal line in the waterfall. The line acts as a border between signals from different minutes.

Table 3.1
Common JT65 and JT9 Frequencies

(All frequencies are in kHz and assume a transceiver operating in USB mode.)

JT65	*JT9*
1838	1840
3576	3578
7076 (European stations often use 7039)	7078
14076	14078
10138	10140
18102	18104
21076	21078
24917	24919
28076	28078
50276	50278

Simultaneous JT65 and JT9: A Potential Transmitting Problem

At the time this book was written, only Dr Joe Taylor's *WSJT-X* software was capable of operating in JT9. As we discussed, one of the neat features of *WSJT-X* is its ability to monitor JT65 and JT9 signals at the same time. It is able to do this because most JT9 signals appear about 2 kHz above the JT65 signals. The signals are easy to identify because JT65 signals sound almost musical and appear as clusters of dots while JT9 signals sound like single tones and appear as lines.

If your transceiver is operating in USB and you set the frequency for, say, 14.076 MHz, the receive audio bandwidth is sufficiently wide that you'll be able to receive all the way to at least 14.078 MHz, if not well beyond, and simultaneously decode JT65 and JT9 signals.

While this is terrific for monitoring, transmitting can present a problem. Let's say you decide to answer a JT9 CQ that appears at the 2700 Hz position in the waterfall. As *WSJT-X* keys your radio and begins transmitting, you may be astonished to see that your RF output is zero!

Why is this happening?

The reason you have zero RF output is because you are attempting to send an audio signal at 2700 Hz to a radio that is running SSB with a typical 2400 Hz *transmit audio bandwidth*. In other words, the frequency of your JT9 audio signal is beyond your transceiver's audio frequency bandwidth for SSB.

Many modern transceivers include the ability to extend their SSB transmit bandwidths, sometimes considerably. If you decide to adjust this parameter, take care. You may experience some unintended consequences, such as distortion. If you enjoy operating SSB voice, you'll have to remember to return the transceiver to its default SSB transmit bandwidth when you've finished running JT65/JT9.

An easier solution is to "split the difference," so to speak. Rather than, for example, setting your transceiver to 14.076 MHz, try setting it to 14.076.5 or even 14.077 MHz. This will allow your transceiver to effectively straddle the JT65 and JT9 areas, bringing both the JT65 and JT9 signals within the 2400 Hz transmit audio bandwidth.

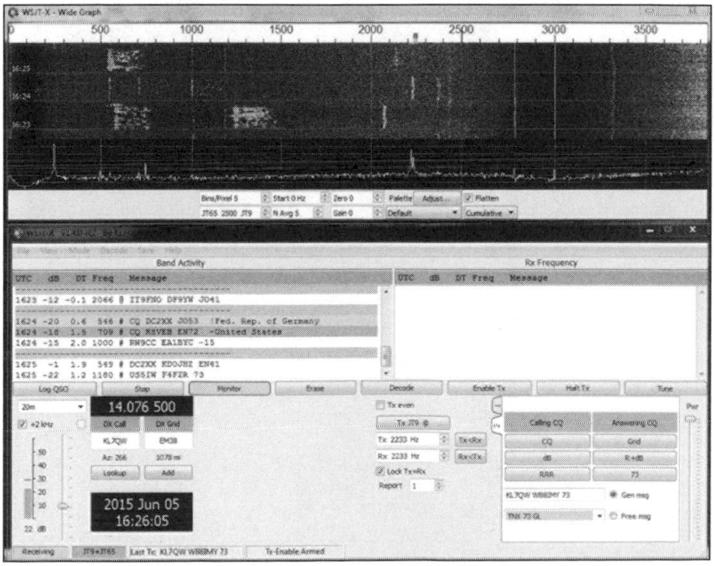

In this screen image, the transceiver's VFO has been set to 14.076.5 MHz. Notice that JT65 and JT9 signals (the solid, wavering lines) both appear in the waterfall *below* the 2400 Hz mark. Even if you attempted to make contact with the highest-frequency JT9 signal shown here, your transceiver should still generate RF output, although possibly at a reduced level.

As you watch JT65 and/or JT9 signals, you'll notice that some are quite faint, just barely visible in the waterfall. At the same time, you may also see signals that appear as broad splotches or smears in the display. Some of these signals are overmodulated; the operator is driving his transceiver with way too much audio. But other signals are simply strong. Even if you are only transmitting a few watts of RF, conditions can exist in which your signal is surprisingly loud at the receiving end.

When the clock reaches 48 seconds, all transmissions will abruptly cease (or at least they should). At this point everyone's software is busy decoding signals. Within a couple of seconds you should see the results.

Most text will be black or gray. These are transmissions between stations engaged in contacts. If you see text with green backgrounds, these are stations calling CQ.

Look at the strength (dB) information for various signals. Depending on the sensitivity of your station and the amount of local noise, you may discover that you've decoded some extremely weak transmissions, perhaps as much as 27 dB below the noise floor. You may even have the spooky experience of decoding a signal on what seems to be an empty "dead" band. That's the power of JT65 and JT9!

Look at the DT (time deviation) information as well. Do most of the signals show DTs of less than 1 second? If so, that's good because it means that your station computer is within fairly close time

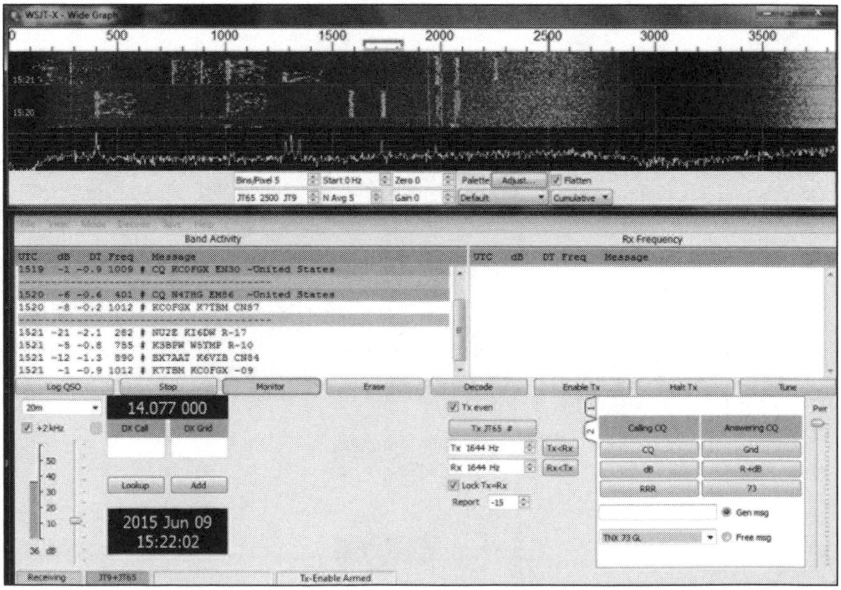

The *WSJT-X* main screen.

synchronization. On the other hand, if most of the decoded signals indicate DTs of *more* than 1 second, your computer may be in need of re-synchronizing.

When the clock ticks off one second into the next minute, the transmissions will begin once again. Soon you will see what the other stations are sending in reply.

Let's Answer a CQ

Once you feel comfortable monitoring activity, why not try answering a CQ? This can be substantially more exciting than it seems, especially when you realize that you have to make choices and click your mouse cursor with just seconds to spare.

Tune to a JT65/JT9 frequency and, if you are using *WSJT-X*, select your mode (JT65 or JT9). Watch the traffic for a few minutes. You're looking for CQ transmissions highlighted in green.

The green lines will appear as soon as your computer decodes the signals, usually at about the 48- to 52-second point. When you see one, you have only *seconds* to double click on the line to reply.

When you double click on the green CQ line, you'll see that the text of your transmission appears automatically in the **TX Generated** or **Gen msg** windows. Don't worry about selecting the even or odd minute; the software "knows" that you want to transmit during the next available minute and it will select even or odd accordingly. If you are using *JT65-HF*, you must also quickly click the **Enable TX** button to "arm" the transmit function.

It is worth mentioning that *WSJT-X* "knows" the mode of the signal you have selected. If you are monitoring JT65 and JT9 at the same time and you click on a CQ from a JT9 station, *WSJT-X* will choose the correct mode automatically and arm itself for transmission.

One second after the beginning of the next minute, when the two right-hand UTC "seconds" digits read "01," the software will key your radio and begin transmitting. This is a good time to check the ALC activity on your radio's meter. If the ALC activity is excessive, reduce the transmit audio from your interface or computer.

The software will continue transmitting your message until just before the 48 second point. There is nothing to do but sit back and wait. At the top of the next minute, a responding station may begin transmitting. About 48 seconds later, you will see his text, which will contain a signal report, highlighted in red. The line is shaded red for a good reason: You need to take action immediately!

Quickly double click on the red line and the software should fill in the text for your next transmission. You'll be sending an "R" to acknowledge his report followed by a report of your own. The software calculates the other station's signal strength and fills in the number (in dB) for you. When the clock reaches one second after the top of the minute, the software will begin sending.

If all goes well, his next transmission will be "RRR" highlighted in red. Double click on this line and the your software will finish the contact by sending 73. If you've activated the CW ID function in your software, it will send your call sign in CW.

Important: Even though your contact has ended, the software may *not* have disabled transmission. If it hasn't, it will repeat your last transmission at the next opportunity. Both *WSJT-X* and the latest version of *JT65-HF* include so-called "watch dog" functions. When enabled, the software will automatically stop transmitting if the same message is sent repeatedly. Make sure this feature is ***on!*** *WSJT-X* has an additional feature to disable transmission after a "73" message is sent. Make sure this is enabled as well.

Call CQ

When you've become comfortable answering a few CQs, it is time to try one of your own.

Let's say you've picked a clear spot in the waterfall display. Click your mouse cursor on that spot and the red/green marker will move to that location. This is where you will be transmitting and receiving.

If you are running *WSJT-X*, select your mode: JT65, JT9 or both. If you intend to operate JT65 and JT9 at the same time, take a moment to read the sidebar "Simultaneous JT65 and JT9: A Potential Transmitting Problem."

When you are ready, click the CQ button. The software responds by generating the appropriate text and setting you up to transmit on the next available minute.

When someone replies to your CQ, you'll see the text highlighted in red as before. Just double click on the red lines and the software will create the proper exchanges automatically. The trick is to stay on your toes and click on the appropriate line before the next minute begins. Otherwise, the software will simply re-send the previous line. The other station will be confused and you'll both have to wait through another set of exchanges to complete the contact.

Once the contact is complete, click the appropriate button to enter the information into the log. You may use separate logging software, but it is a good idea to log it here as well, just in case.

Chapter 4

Tips and Tricks

When it comes to JT65 and JT9, or any other mode of operating for that matter, there is no substitute for experience. The more you get on the air, the more you will learn.

That said, I've included this brief chapter with the hope of enhancing your enjoyment of JT65 and JT9 by helping you avoid some of the common pitfalls, and by passing along some ideas that may expand the usefulness of these modes.

First, the pitfalls.

Dirty Digital

In Chapter 2 we discussed the need to maintain your transmit audio at levels that will result in clean signals. Let's spend some time going into a little more detail so that you have a clearer understanding of the problem, not just for JT65 or JT9, but for all HF digital modes.

Not only is it extremely important to do all you can to ensure that you are transmitting clean signals, it is equally important that you not accidentally accuse another operator of being "dirty" when, in fact, his signals are quite clean. Yes, this is possible and it happens more often than you may realize.

Computer sound devices, and even the sound chips within digital mode interfaces, are capable of producing substantial amounts of transmit audio – much more than your transceiver can possibly use. To make matters potentially worse, some amateurs are enjoying digital modes with computers that only offer speaker or headphone audio outputs – signals powerful enough to be deafening when applied directly to one's ears, let alone to delicate transceiver audio stages.

A transceiver's microphone input is designed for exquisitely weak signals; it was never intended to be a conduit for higher-level audio. When you apply excessive audio to microphone amplification circuitry, you run the risk of driving those circuits into non-linear operation, turning them into prodigious distortion generators. The distortion products they create are passed through the amplification and frequency conversion stages until they eventually become horribly distorted RF signals at the output.

This is not to say that you should never apply digital signal audio to a transceiver's microphone jack. You certainly can – if you are careful to keep the audio signal at a very low level. As we discussed previously, one easy way to do this is to watch your transceiver's ALC activity. Most SSB transceivers display ALC performance, typically with a mechanical meter or an LCD or LED indicator. With your rig in the transmit mode, slowly increase the audio output of your computer or interface until you begin to see ALC activity.

Depending on the design of your meter, you may see the needle or the LED/LCD display suddenly rise from zero. Other transceiver meters display an ALC "safe" zone in which the needle or LED/LCD indicator fluctuates within what is considered an acceptable range.

With the Elecraft K3s and KX3s, the multi-segmented meter marked "ALC" works in a somewhat different fashion. The first four segments of the meter can be considered to be an indication of the audio drive level, and is quite useful for that purpose. It is when the *fifth* segment illuminates that you need to be concerned. That's when the ALC is becoming active. So, with the K3 or KX3, you should decrease the audio drive until the fifth segment barely flickers.

Regardless of the how the meter operates, the goal is to keep ALC activity as low as possible (or at least keep it from exceeding the acceptable range). In my Kenwood TS-2000 transceiver, I monitor the ALC with the goal of adjusting audio levels for no ALC activity whatsoever.

The Accessory Jack Alternative

As we also discussed earlier, most modern rigs have a rear panel accessory jack with pins for all sorts of functions, including audio inputs and outputs and PTT keying. This is, by far, the best jack to use for digital audio interfacing. The audio output is usually at a fixed level, which is great for receiving. Better yet, the audio input is designed to accommodate higher signal levels. You can still overdrive your transmit audio at the accessory jack, but not quite as easily.

Three Rules

When operating sound-device-based digital modes, regardless of where you apply the audio, here are three rules to help keep you on the straight and narrow:

• Monitor your transceiver ALC and adjust the transmit audio from your computer or interface for minimal ALC activity.

• Use the minimum amount of transmit audio necessary to achieve the desired RF output power. In addition, avoid the trap of running your computer or interface transmit audio at high levels and then trying to compensate by reducing your transceiver's microphone gain (if you are using the microphone input) to keep your ALC in check. You may think this will guarantee a clean signal, but there is a good chance that by firewalling your transmit audio you are creating a dirty signal before it even reaches your radio. Start with low audio levels from your computer or interface.

• If your transceiver lacks an ALC display, increase your computer or interface audio until the RF output level stops rising, and then *reduce* the audio level until the RF output declines by about 50%. (Trust me: You won't miss the 50%.)

None of the preceding steps will absolutely guarantee a clean signal, but they'll tilt the odds strongly in your favor.

When Digital Isn't Dirty After All

With the proliferation of waterfall displays, and a lack of understanding about how they work, we've seen the rise of the *Dirty Digital Police*. These are hams who've made it their missions to rid the HF digital world of overmodulated signals. Their hearts are in the right places, but in some instances they have identified dirty signals (and harangued allegedly guilty operators) when, in fact, the offending signals were quite clean.

How can this be? If you see a signal smearing halfway across your JT65 waterfall display, it must be dirty, right? Not necessarily. There is a good chance the "dirty operator" is being falsely accused. Here's why.

Most waterfall displays are intended to be tuning indicators. They provide a visual representation of a digital signal so that you can use your software to "lock onto" the signal and decode it. These waterfall displays are not intended to be spectrum analyzers – especially when signals are strong.

And therein lurks the problem. At least two things occur in the presence of a strong signal: (1) The transceiver's Automatic Gain Control (AGC) will instantly reduce receiver gain to maintain linearity. The result is that weaker signals will fade on the waterfall or disappear entirely. (2) The sound circuitry in your computer or interface will be swamped with audio, causing it to exhibit non-linear behavior. The result is the distortion you see on your waterfall.

Before you declare a signal to be irredeemably dirty, reduce the RF gain of your receiver. Most hams run their rigs with their RF gain controls set to maximum. This is unnecessary and, with digital modes at least, extremely deceiving. Turn down the RF gain (or switch in some attenuation) and keep turning it down as you watch the waterfall display. Does the offending signal suddenly become much cleaner in the display? If so, that should tell you something.

Another trick is to reduce the receive audio level between your radio and your sound card. Some interfaces make this easy with audio level knobs on their front panels. Once again, as you reduce the audio you may find that dirty signals become magically clean. If so, the signals were never dirty to begin with.

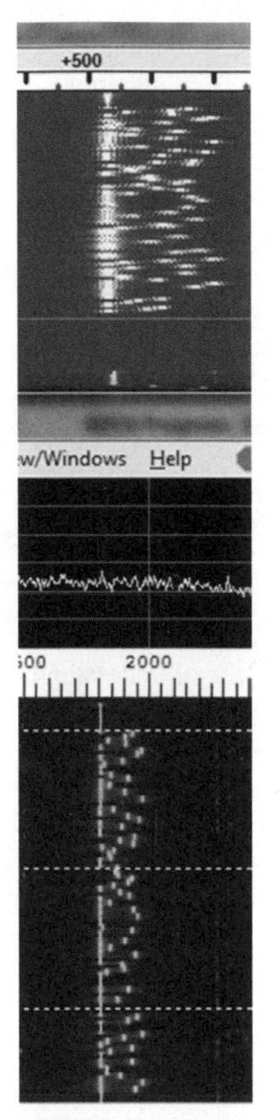

Figure 4.1 – A comparison between a JT65 signal received in a *JT65-HF* waterfall (top) and *Spectrum Lab* (bottom). The signal looks somewhat smeared and distorted in the *JT65-HF* waterfall, which might lead you to believe that the transmitting station was overmodulating. However, in *Spectrum Lab* the same signal is quite clean.

Figure 4.1 offers an interesting comparison. At the top you'll see a JT65 signal as received by *JT65-HF* software. This is a strong signal and it certainly looks distorted, doesn't it? Notice how it seems to smear within the waterfall. But look at the lower portion of the Figure 5 image. This is the same signal, received at the same time, using *Spectrum Lab* software (**www.qsl.net/dl4yhf/spectra1.html**). Unlike the *JT65-HF* waterfall display, *Spectrum Lab* is specifically designed to analyze signal characteristics. Notice how the so-called "dirty" JT65 signal is, in fact, acceptably clean in *Spectrum Lab*.

When You Find A Dirty Signal . . .

Before you declare a signal to be dirty, follow the steps in this article and make sure that what you are seeing is *truly* a dirty signal. If the signal is indeed dirty, do not humiliate the operator on the air by transmitting DIRTY SIGNAL or something equally rude. Chances are the other guy has no idea what is going on. Your nasty comment won't be helpful; it will only serve to cause anger and confusion.

Instead, press and hold the SHIFT key on your computer keyboard and then tap the PRT SCN (Print Screen) key. You've just temporarily captured the screen image in memory. Now open an image processing or graphics program such as *Irfanview* (**www.irfanview.com**) or Microsoft *Paint*. Paste the stored image into the program and save it to your hard drive.

Next, hunt down the e-mail address of the operator. Try QRZ.com. If he is an ARRL member, he may also have an ARRL e-mail address such as **callsign@arrl.net**.

Send the signal image along with a polite note indicating the time and frequency. Offer suggestions about how he might try reducing his transmit audio, etc. Avoid sounding hostile or condescending. End the message with the hope that you'll hear him on the air in the future.

"Why Can't I Decode Signals?"

This is possibly the most common question I receive from new JT65 and JT9 operators. They've installed their software and they are seeing evidence of signals in their waterfall displays, yet they never (or rarely) see decoded information.

If you can see signals in the waterfall, but nothing decodes, the usual culprit is poor time synchronization. We discussed this issue earlier in the book, but it deserves emphasizing.

Both JT65 and JT9 depend on relatively tight time synchronization to work their magic. If your station computer is more than two seconds out of sync with the computer at the transmitting station, you will not decode signals. (For JT9 the sync tolerance is only one second.) Follow the techniques described in Chapter 3 to ensure that your computer's clock is set as accurately as possible.

In most cases an accurate clock setup will solve the problem. However, there are other, less common, issues that can result in failure to decode.

Once you chose a band on which to operate, park your radio on the frequency of your choice *and then don't touch the VFO again*. The same goes for the RIT or Clarifier controls. You don't want to touch any controls that will change the frequency of the radio. Even a tiny frequency change is deadly to the ability to decode JT65 or JT9. And as I mentioned earlier in the book, if you are using an older radio, or a radio that tends to change frequency while warming up, let the rig warm up and stabilize for about 20 to 30 minutes before attempting to operate.

It goes without saying – because we've discussed it already – but I'd be remiss if I didn't emphasize the need to set your receive audio to a reasonable level. Audio level adjustment doesn't have to be precise, but you need to give the software enough audio to work with. By the same token, you don't want to hammer your computer with excessive audio. Beyond a certain point, turning up your receive audio will not improve your ability to decode signals. In fact, too much audio will result in few, if any, decoded signals.

Finally, although it isn't a common problem, it is possible that you may have a program in the background that is interfering with your JT65/JT9 software. If you notice that *WSJT-X* or *JT65-HF* are abnormally slow to start, it may be time to investigate a possible software conflict. Shut down and re-boot your computer. This may clear the offending program from memory and solve the problem. If not, open *Windows Task Manager* and see what other applications may be running in the background.

Reverse Beacons

One of the most fascinating aspects of the JT65/JT9 world is the way in which it has harnessed the Internet as a powerful tool to share information. It is fair to say that the Internet functions almost as a kind of "second ionosphere" for JT65 and JT9 operators.

It may surprise you to know that whenever you transmit, your signal is being heard by more stations than just the ones you see in your waterfall display. There are many stations (the exact number varies by frequency) that are silently monitoring without transmitting. Not only are these stations not transmitting, they may be completely empty with no human operators in sight.

All of the information that is gathered by these monitoring stations is sent, via the Internet, to websites that function as *reverse beacons*. Unlike traditional beacons that radiate signals 24/7 so that hams can keep an eye on ever-changing propagation conditions, a reserve beacon collects information and then makes it available to everyone via designated websites. Think of a reserve beacon as a kind of information vacuum cleaner, sucking up data and holding it in one place.

JT65-HF and *WSJT-X* both have the ability to automatically upload reports to reverse beacons. As a result, JT65 and JT9 are among the most popular modes in the reverse beacon community. At any given time where are thousands of JT65 and JT9 stations monitoring and uploading to reverse beacon sites. They are uploading the call signs of the stations they "hear" in their waterfalls, plus the received signal strengths in dB. The uploads are continuous, minute by minute, and the websites also update their displayed information each minute.

The result is a wealth of data that hams can access anytime of the day or night. You can instantly see who is receiving your signals and how strong your signals are at their locations. You can also see which stations are monitoring in various countries, and which frequencies they are monitoring. Some stations monitor only single frequencies, but others eavesdrop on multiple frequencies simultaneously (often using Software Defined Radios).

If you have Internet connectivity at your station, I strongly encourage you to enable uploading to PSKReporter and other reverse beacons. Uploading will not affect the performance of your station in any way, nor does it require any action on your part. Everything happens automatically in the background whenever your JT65/JT9 software is running.

I sometimes start my JT65 software, turn on my radio, and then leave to do household chores or whatever else is on my agenda for the day. My station will sit unattended for hours, listening to the frequency I've chosen and uploading the results to the reverse beacon networks. It costs me nothing (unless you count the electricity usage) and it benefits a very large audience.

It is important to note that these reverse beacons are strictly volunteer efforts. They are free for you to use, but bear in mind that someone is paying for the software, Internet access and web hosting. That "someone" may be an individual or a club. Either way, you may want to consider sending an occasional monetary donation to help defray the expenses of these valuable services.

PSKReporter

Philip Gladstone, N1DQ, is the creator and administrator of PSKReporter at **http://pskreporter.info/**. Originally intended as a tool to track PSK31 activity, PSKReporter has blossomed into a multimode reverse beacon that reports everything from CW to JT9. Of particular interest is its map display at **https://www.pskreporter.info/pskmap.html**. At this page (**Figure 4.2**) you can see a graphic presentation of every monitoring station that has received your signals on any band. When you move

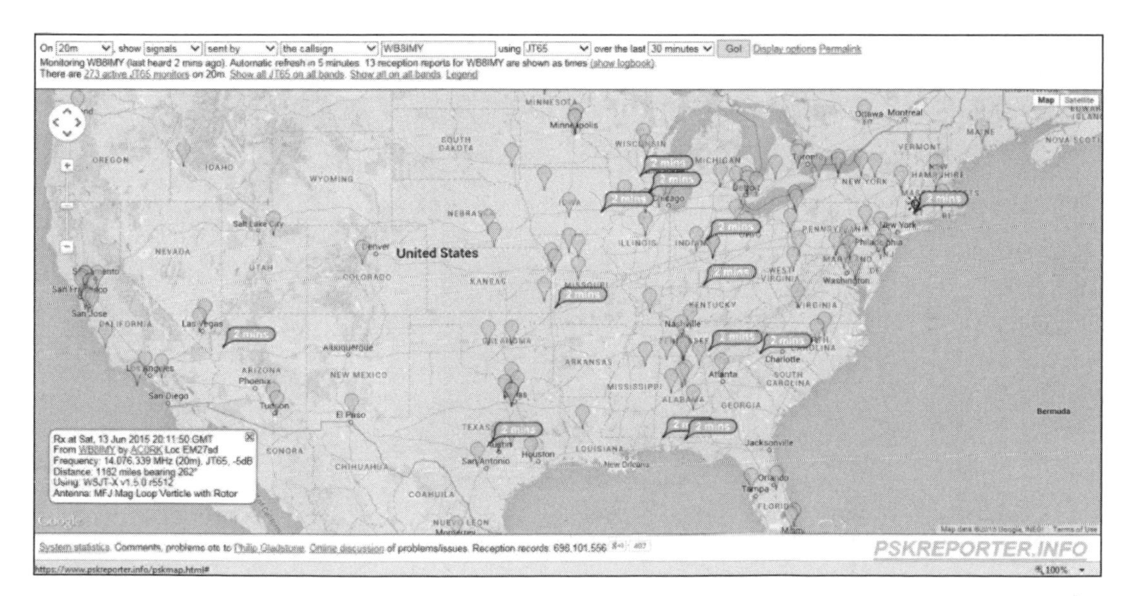

Figure 4.2 – PSKReporter showing the results of one of my JT65 transmission on 20 meters that took place just two minutes before this image was captured. Each "2 mins" balloon represents a station that received my transmission. In the lower left corner you can see the result of my mouse cursor touching on the balloon that marks ACØRK's station. A report of -5 dB from a station using a magnetic loop antenna is impressive.

your mouse cursor over a monitoring station's "balloon," a box will open to provide more information about the station, often including a signal report.

A number of monitoring stations also include brief descriptions of the types of antennas they are using. This is helpful when you are trying to get a better sense of how well your signal is being heard. After all, a strong -4 dB report from a station using a wire dipole inside an attic is much more impressive that the same report from a station using a Yagi antenna installed on a 100-foot tower!

Hamspots

Hamspots (**http://hamspots.net/**) is a reverse beacon system developed by Laurie Cowcher, VK3AMA (see **Figure 4.3**). Rather than attempting to aggregate data about all modes, Hamspots concentrates on digital modes. Hamspots lists reports from monitoring stations, but also includes reports of digital activity that have been placed on popular DX Clusters. You can search for reports of your call sign, or any other call sign. You can even access a real-time chat function to exchange comments. The chat window is useful when you are doing research and want to set up a test with another station.

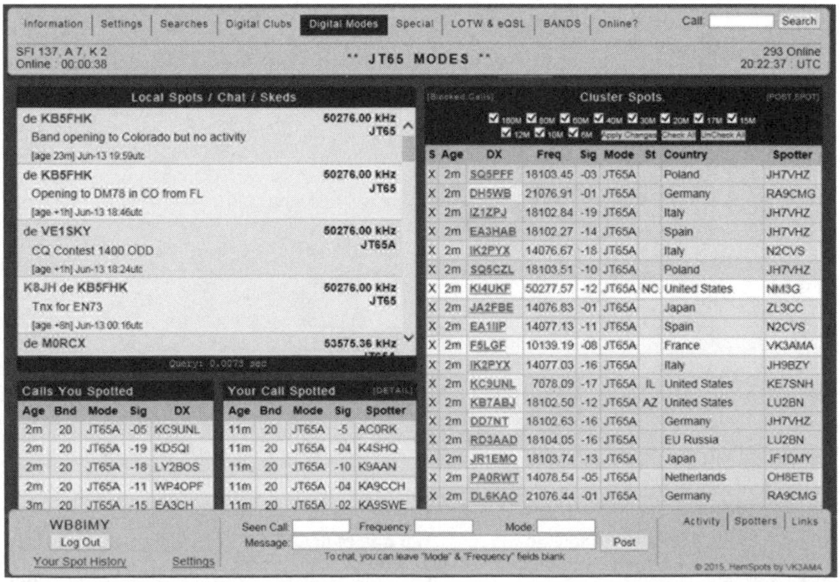

Figure 4.3 – The Hamspots reverse beacon.

Reverse Beacon Research

While it is great fun to see who is hearing your JT65 or JT9 transmissions, it is even more interesting to use that information to get a better sense of how your station is really performing.

To offer a personal example, I enjoy experimenting with antennas, but I live on a very small lot, so the challenge is to maximize performance within a limited space. Recently, I ran some comparison tests between my trusty multiband vertical antenna and a new inverted V antenna.

I used an antenna switch to select either the vertical or the inverted V. Sitting down at the radio at 2049 UTC, I selected the inverted V antenna and set the transceiver VFO to 14.076 MHz. I noted the time on a piece of scrap paper, along with my RF power output, and then sent a JT65 CQ. At 2051 UTC, I sent another CQ, but this time I selected the vertical antenna. In **Figure 4.4** you can see the result of my query to the Hamspots reverse beacon about 10 minutes later.

25	P	Jun-13	20:51	8 mins	KM4SFF	United States	FL	20	-15	JT65A
26	A	Jun-13	20:51	8 mins	W6IR	United States	IA	20	-10	JT65A
27	X	Jun-13	20:51	8 mins	KL7QW	United States	MO	20	-16	JT65A
28	X	Jun-13	20:51	8 mins	ND4Q	United States	AL	20	-19	JT65A
29	X	Jun-13	20:51	8 mins	KF4RWA	United States	AL	20	-10	JT65A
30	X	Jun-13	20:51	8 mins	KK4A	United States	AL	20	-15	JT65A
31	X	Jun-13	20:51	8 mins	K9AAN	United States	KY	20	-17	JT65A
32	X	Jun-13	20:51	8 mins	K4SHQ	United States	AL	20	-07	JT65A
33	X	Jun-13	20:51	8 mins	KB2OBQ	United States	FL	20	-14	JT65A
34	P	Jun-13	20:49	10 mins	KC6ZZG	United States	CA	20	-17	JT65A
35	P	Jun-13	20:49	10 mins	KK6AYK	United States	CA	20	-17	JT65A
36	P	Jun-13	20:49	10 mins	K5RHD	United States	NM	20	-13	JT65A
37	P	Jun-13	20:49	10 mins	W0RSB	United States	MN	20	-8	JT65A
38	X	Jun-13	20:49	10 mins	K9AAN	United States	KY	20	-10	JT65A
39	X	Jun-13	20:49	10 mins	K9TAD	United States	CA	20	-15	JT65A
40	X	Jun-13	20:49	10 mins	VA3TX	Canada		20	-20	JT65A
41	X	Jun-13	20:49	10 mins	KA9SWE	United States	WI	20	-10	JT65A
42	X	Jun-13	20:49	10 mins	KF4RWA	United States	AL	20	-01	JT65A
43	X	Jun-13	20:49	10 mins	KB2OBQ	United States	FL	20	-05	JT65A
44	X	Jun-13	20:49	10 mins	KA8GBB	United States	MI	20	-21	JT65A
45	X	Jun-13	20:49	10 mins	K4SHQ	United States	AL	20	-01	JT65A
46	A	Jun-13	20:49	10 mins	W6IR	United States	IA	20	-07	JT65A
47	X	Jun-13	20:49	10 mins	ND4Q	United States	AL	20	-08	JT65A

Figure 4.4 – Using Hamspots to experiment with higher and lower power transmissions on 20 meters. All of the listed stations received my transmissions at various times.

Notice that KF4RWA in Alabama reported my 2049 UTC transmission on the inverted V as being -01 dB, which is quite strong. However, when I switched to the vertical antenna for the 2051 CQ transmission, he reported my signal as being -10 dB. That is a substantial drop! On the other hand, W6IR in Iowa reported a decline from -7 to -10 dB. A loss of 3 dB is a significant decrease, but not as serious as the report from KF4RWA. It looks like my inverted V is outperforming my vertical – at least on 20 meters.

Of course, anyone with even a slight amount of science training will be quick to point out that these results are not scientifically rigorous. This is true. There are a number of variables in play that could account for some of the differences I am seeing. Maybe W6IR had a Yagi antenna pointed right at me, while KF4RWA was using only a low dipole antenna. Local noise levels can be strong factors, as can variations in the characteristics of the ionosphere.

Even so, if you have the patience to make more measurements at various times, a pattern will start to emerge. For instance, I discovered that my vertical antenna has a lobe in its radiation pattern on 15 meters that points to the southwest. Reports from stations on 15 meters in that part of the United States and beyond are consistently stronger than those from any

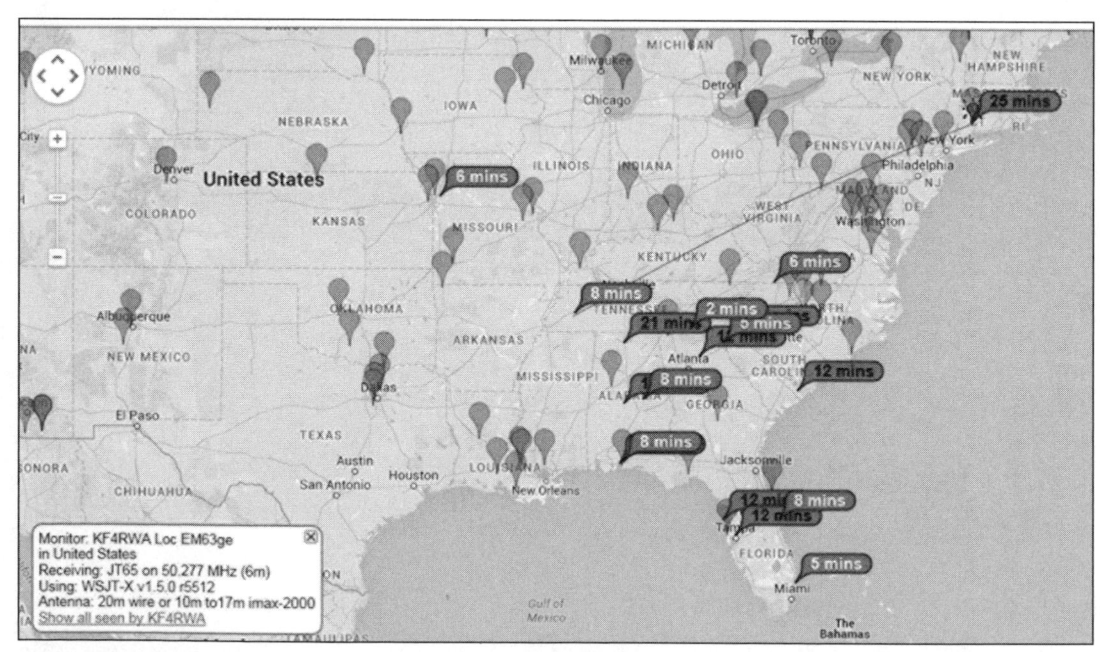

Here is what a 6-meter Sporadic E opening looks like with PSKReporter. JT65 transmissions from my station (the "25 mins" balloon in Connecticut) were heard by several stations in the southeastern United States, and one in western Missouri. Looking at the results, it is likely the Sporadic E "cloud" that reflected my signals was somewhere over West Virginia, Virginia and North Carolina.

other point on the compass. When I switch to the inverted V antenna, the gain the vertical offered in that direction clearly disappears.

It is also fascinating to use the reverse beacons to observe how propagation changes over time. As nightfall approaches, I may spend the evening making contacts on 80 meters and watching the beacon reports as the day/night terminator moves west of my location. Reports from stations in the Midwest will gradually strengthen and reports from stations in the Rockies will begin to appear.

If you have 6 meter capability, JT65 can be particularly interesting. On a number of occasions I have parked my rig on 50.276 MHz and allowed it to monitor the frequency throughout the day. Most of the time, my software fails to decode a single transmission, but every so often it will decode a distant CQ that appears only once before vanishing like a radio ghost.

You can have some fun playing with RF power levels as well. Trying making a few contacts at, say, 40 W output and then drop your power down to 5 W. Examine the reverse beacon data and the results may surprise you! I've seen reports of my signals when I was transmitting at just 500 *milliwatts*.

The Paper Chase

JT65 and JT9 have opened award hunting to large numbers of hams who could not otherwise participate. Already we're seeing hams with indoor

Figure 4.5 – The ARRL Worked All States award.

antennas working 100 entities to earn their initial ARRL DX Century Club (DXCC) certificates; many have also earned DXCC endorsements. ARRL Worked All States awards (**Figure 4.5**) are now routinely issued to operators who've earned their certificates strictly with JT65 and/or JT9 contacts.

Both *JT65-HF* and *WSJT-X* have logging functions. You can export this information to whatever other software you are using and then upload to ARRL's Logbook of The World system. Or, you can simply type entries into your preferred logging program manually as you make your JT65 or JT9 contacts. For example, I often run my logging software in a separate window and then just enter the data "by hand" whenever I complete a contact.

As JT65 and JT9 award hunting became a popular pursuit, it wasn't long before someone developed a software add-on to enhance the game. That "someone" was Laurie Cowcher, VK3AMA. He developed the free *JTAlert* application for both *JT65-HF* (and applications based on it) as well as *WSJT-X*. You can download it from **http://hamapps.com/**. See **Figures 4.6** and **4.7**.

JTAlert works with your JT65/JT9 software to quickly identify stations that you may need for awards such as DXCC or Worked All States. It does this by scanning your *JT65-HF* or *WSJT-X* log to determine whether you have just decoded a signal from a station you need for your Worked All States, DXCC, or other awards.

If you have decoded a CQ from a station you need, *JTAlert* will let you know with visible and audible alerts. The software can provide alerts for:

- Your call sign decoded (someone calling you)
- CQ and QRZ
- A wanted call sign
- A wanted Prefix (by band/mode)
- A wanted Grid (by band/mode)
- A wanted US State (by band/mode)
- A wanted DXCC (by band/mode)
- A wanted CQ Zone (by band/mode)
- A wanted Continent (by band/mode)

Let's say you are pursuing your ARRL Worked All States award. You're monitoring JT65 activity while otherwise puttering around in your station.

Suddenly, you hear "Wanted state!" from your computer speakers.

You look at the JTAlert display and see that N6ATQ in California has called CQ. You haven't worked California on JT65 and this fact has caused *JTAlert* to highlight this call sign in yellow with red letters to let

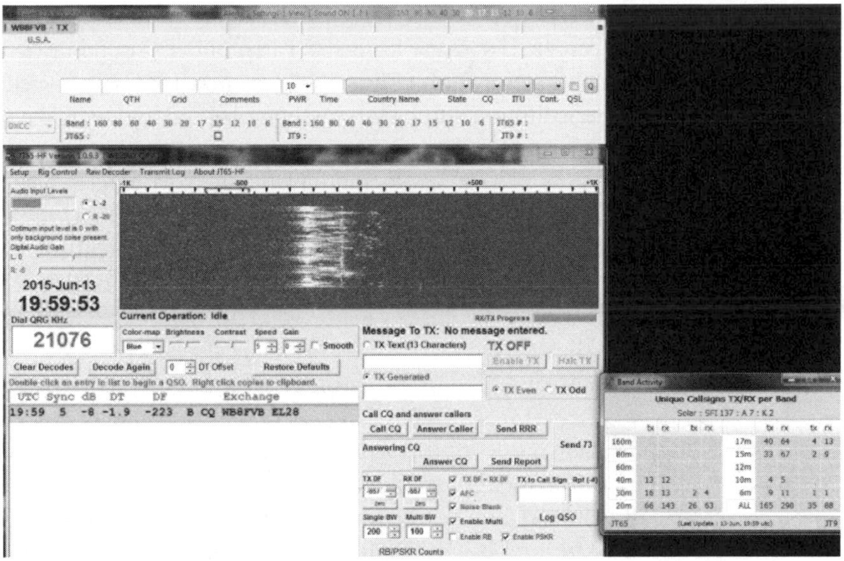

	tx	rx	tx	rx		tx	rx	tx	rx
160m					17m	40	64	4	13
80m					15m	33	67	2	9
60m					12m				
40m	13	12			10m	4	5		
30m	16	13	2	4	6m	9	11	1	1
20m	66	143	26	63	ALL	165	290	35	88

Figure 4.6 – Running *JTAlert* with *JT65-HF*.

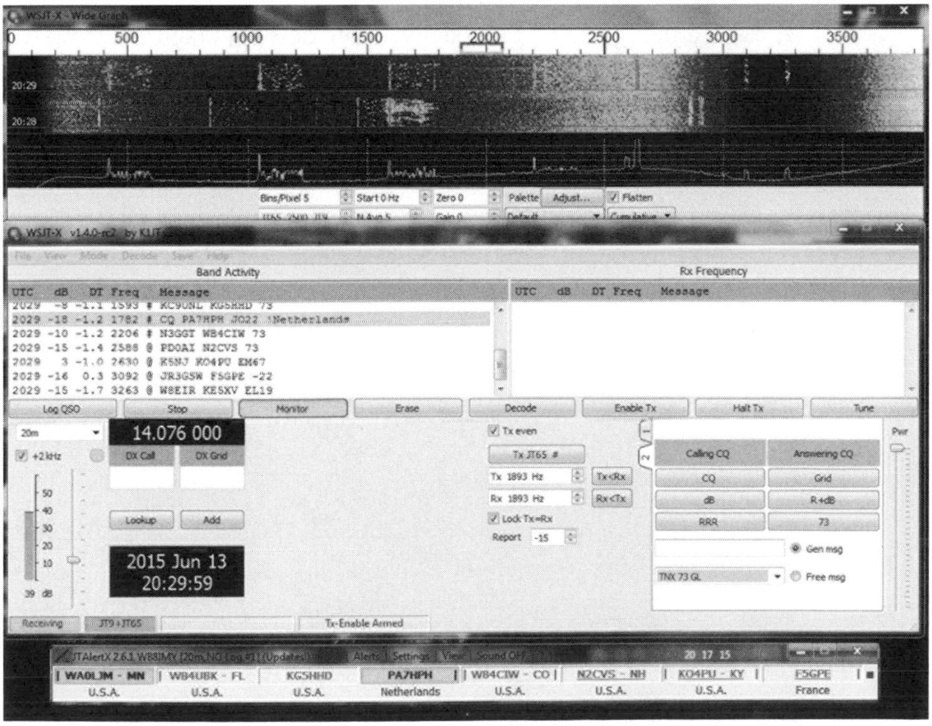

Figure 4.7 – *JTAlert* and *WSJT-X*. In this configuration, *JTAlert* is displaying the list of received stations and their locations below the *WSJT-X* primary window.

you know that you need to answer the CQ right away!

Of course, if you are using your computer's sound device to encode and decode signals from your radio, you won't be able to enjoy the audio alerts. The solution is to install a second sound device in your PC, or use an interface that has its own sound device built in.

I find the *JTAlert* audio announcements particularly useful if I happen to become distracted during a contact. My cat might jump into my lap and demand a thorough head scratching while I am waiting for signals to decode. Fortunately, *JTAlert* will draw my attention back to the monitor with an insistent "Calling you!" as soon as the other station's transmission has been decoded.

In *JTAlert* will perform automatic logging if you are using *DX-Keeper*, *Log4OM*, *HRD Log V5/6* or *MixW*. It will also upload to online logbooks such as ClubLog.org and HRDLog.net. All stations you decode will be automatically uploaded to the HamSpots reverse beacon.

It is interesting to note that *JTAlert* will flag a station if you have worked it before. It will display the call sign against a gray background with the text "B4." There are some JT65 and JT9 operators who will only work a station once, but this is a controversial practice. While I can understand the eagerness to accumulate contacts in pursuit of an award, in my opinion there is always time to spare for a second, third, fourth contact, or more. I don't mind spending a measly five minutes out of my life to work someone again.

Index